T0192393

Statistical Topics in Health Economics and Outcomes Research

Chapman & Hall/CRC Biostatistics Series

Editor-in-Chief

Shein-Chung Chow, Ph.D., Associate Director, Office of Biostatistics, CDER/FDA, Silver Springs, Maryland

Series Editors

Byron Jones, Biometrical Fellow, Statistical Methodology, Integrated Information Sciences, Novartis Pharma AG, Basel, Switzerland

Jen-pei Liu, Professor, Division of Biometry, Department of Agronomy, National Taiwan University, Taipei, Taiwan

Karl E. Peace, Georgia Cancer Coalition, Distinguished Cancer Scholar, Senior Research Scientist and Professor of Biostatistics, Jiann-Ping Hsu College of Public Health, Georgia Southern University, Statesboro, Georgia

Bruce W. Turnbull, Professor, School of Operations Research and Industrial Engineering, Cornell University, Ithaca, New York

Published Titles

Adaptive Design Methods in Clinical Trials, Second Edition
Shein-Chung Chow and Mark Chang

Adaptive Designs for Sequential Treatment Allocation
Alessandro Baldi Antognini and Alessandra Giovagnoli

Adaptive Design Theory and Implementation Using SAS and R, Second Edition
Mark Chang

Advanced Bayesian Methods for Medical Test Accuracy
Lyle D. Broemeling

Analyzing Longitudinal Clinical Trial Data: A Practical Guide
Craig Mallinckrodt and Ilya Lipkovich

Applied Biclustering Methods for Big and High-Dimensional Data Using R
Adetayo Kasim, Ziv Shkedy, Sebastian Kaiser, Sepp Hochreiter, and Willem Talloen

Applied Meta-Analysis with R
Ding-Geng (Din) Chen and Karl E. Peace

Applied Surrogate Endpoint Evaluation Methods with SAS and R
Ariel Alonso, Theophile Bigirumurame, Tomasz Burzykowski, Marc Buyse, Geert Molenberghs, Leacky Muchene, Nolen Joy Perualila, Ziv Shkedy, and Wim Van der Elst

Basic Statistics and Pharmaceutical Statistical Applications, Second Edition
James E. De Muth

Bayesian Adaptive Methods for Clinical Trials
Scott M. Berry, Bradley P. Carlin, J. Jack Lee, and Peter Muller

Bayesian Analysis Made Simple: An Excel GUI for WinBUGS
Phil Woodward

Bayesian Designs for Phase I–II Clinical Trials
Ying Yuan, Hoang Q. Nguyen, and Peter F. Thall

Bayesian Methods for Measures of Agreement
Lyle D. Broemeling

Bayesian Methods for Repeated Measures
Lyle D. Broemeling

Bayesian Methods in Epidemiology
Lyle D. Broemeling

Bayesian Methods in Health Economics
Gianluca Baio

Bayesian Missing Data Problems: EM, Data Augmentation and Noniterative Computation
Ming T. Tan, Guo-Liang Tian, and Kai Wang Ng

Bayesian Modeling in Bioinformatics
Dipak K. Dey, Samiran Ghosh, and Bani K. Mallick

Benefit-Risk Assessment in Pharmaceutical Research and Development
Andreas Sashegyi, James Felli, and Rebecca Noel

Benefit-Risk Assessment Methods in Medical Product Development: Bridging Qualitative and Quantitative Assessments
Qi Jiang and Weili He

Bioequivalence and Statistics in Clinical Pharmacology, Second Edition
Scott Patterson and Byron Jones

Biosimilar Clinical Development: Scientific Considerations and New Methodologies
Kerry B. Barker, Sandeep M. Menon, Ralph B. D'Agostino, Sr., Siyan Xu, and Bo Jin

Biosimilars: Design and Analysis of Follow-on Biologics
Shein-Chung Chow

Biostatistics: A Computing Approach
Stewart J. Anderson

Cancer Clinical Trials: Current and Controversial Issues in Design and Analysis
Stephen L. George, Xiaofei Wang, and Herbert Pang

Causal Analysis in Biomedicine and Epidemiology: Based on Minimal Sufficient Causation
Mikel Aickin

Published Titles

Clinical and Statistical Considerations in Personalized Medicine
Claudio Carini, Sandeep Menon, and Mark Chang

Clinical Trial Data Analysis Using R
Ding-Geng (Din) Chen and Karl E. Peace

Clinical Trial Data Analysis Using R and SAS, Second Edition
Ding-Geng (Din) Chen, Karl E. Peace, and Pinggao Zhang

Clinical Trial Methodology
Karl E. Peace and Ding-Geng (Din) Chen

Clinical Trial Optimization Using R
Alex Dmitrienko and Erik Pulkstenis

Cluster Randomised Trials: Second Edition
Richard J. Hayes and Lawrence H. Moulton

Computational Methods in Biomedical Research
Ravindra Khattree and Dayanand N. Naik

Computational Pharmacokinetics
Anders Källén

Confidence Intervals for Proportions and Related Measures of Effect Size
Robert G. Newcombe

Controversial Statistical Issues in Clinical Trials
Shein-Chung Chow

Data Analysis with Competing Risks and Intermediate States
Ronald B. Geskus

Data and Safety Monitoring Committees in Clinical Trials, Second Edition
Jay Herson

Design and Analysis of Animal Studies in Pharmaceutical Development
Shein-Chung Chow and Jen-pei Liu

Design and Analysis of Bioavailability and Bioequivalence Studies, Third Edition
Shein-Chung Chow and Jen-pei Liu

Design and Analysis of Bridging Studies
Jen-pei Liu, Shein-Chung Chow, and Chin-Fu Hsiao

Design & Analysis of Clinical Trials for Economic Evaluation & Reimbursement: An Applied Approach Using SAS & STATA
Iftekhar Khan

Design and Analysis of Clinical Trials for Predictive Medicine
Shigeyuki Matsui, Marc Buyse, and Richard Simon

Design and Analysis of Clinical Trials with Time-to-Event Endpoints
Karl E. Peace

Design and Analysis of Non-Inferiority Trials
Mark D. Rothmann, Brian L. Wiens, and Ivan S. F. Chan

Difference Equations with Public Health Applications
Lemuel A. Moyé and Asha Seth Kapadia

DNA Methylation Microarrays: Experimental Design and Statistical Analysis
Sun-Chong Wang and Arturas Petronis

DNA Microarrays and Related Genomics Techniques: Design, Analysis, and Interpretation of Experiments
David B. Allison, Grier P. Page, T. Mark Beasley, and Jode W. Edwards

Dose Finding by the Continual Reassessment Method
Ying Kuen Cheung

Dynamical Biostatistical Models
Daniel Commenges and Hélène Jacqmin-Gadda

Elementary Bayesian Biostatistics
Lemuel A. Moyé

Emerging Non-Clinical Biostatistics in Biopharmaceutical Development and Manufacturing
Harry Yang

Empirical Likelihood Method in Survival Analysis
Mai Zhou

Essentials of a Successful Biostatistical Collaboration
Arul Earnest

Exposure–Response Modeling: Methods and Practical Implementation
Jixian Wang

Frailty Models in Survival Analysis
Andreas Wienke

Fundamental Concepts for New Clinical Trialists
Scott Evans and Naitee Ting

Generalized Linear Models: A Bayesian Perspective
Dipak K. Dey, Sujit K. Ghosh, and Bani K. Mallick

Handbook of Regression and Modeling: Applications for the Clinical and Pharmaceutical Industries
Daryl S. Paulson

Inference Principles for Biostatisticians
Ian C. Marschner

Interval-Censored Time-to-Event Data: Methods and Applications
Ding-Geng (Din) Chen, Jianguo Sun, and Karl E. Peace

Introductory Adaptive Trial Designs: A Practical Guide with R
Mark Chang

Joint Models for Longitudinal and Time-to-Event Data: With Applications in R
Dimitris Rizopoulos

Measures of Interobserver Agreement and Reliability, Second Edition
Mohamed M. Shoukri

Medical Biostatistics, Fourth Edition
A. Indrayan

Meta-Analysis in Medicine and Health Policy
Dalene Stangl and Donald A. Berry

Methods in Comparative Effectiveness Research
Constantine Gatsonis and Sally C. Morton

Mixed Effects Models for the Population Approach: Models, Tasks, Methods and Tools
Marc Lavielle

Modeling to Inform Infectious Disease Control
Niels G. Becker

Published Titles

Modern Adaptive Randomized Clinical Trials: Statistical and Practical Aspects
Oleksandr Sverdlov

Monte Carlo Simulation for the Pharmaceutical Industry: Concepts, Algorithms, and Case Studies
Mark Chang

Multiregional Clinical Trials for Simultaneous Global New Drug Development
Joshua Chen and Hui Quan

Multiple Testing Problems in Pharmaceutical Statistics
Alex Dmitrienko, Ajit C. Tamhane, and Frank Bretz

Noninferiority Testing in Clinical Trials: Issues and Challenges
Tie-Hua Ng

Optimal Design for Nonlinear Response Models
Valerii V. Fedorov and Sergei L. Leonov

Patient-Reported Outcomes: Measurement, Implementation and Interpretation
Joseph C. Cappelleri, Kelly H. Zou, Andrew G. Bushmakin, Jose Ma. J. Alvir, Demissie Alemayehu, and Tara Symonds

Quantitative Evaluation of Safety in Drug Development: Design, Analysis and Reporting
Qi Jiang and H. Amy Xia

Quantitative Methods for HIV/AIDS Research
Cliburn Chan, Michael G. Hudgens, and Shein-Chung Chow

Quantitative Methods for Traditional Chinese Medicine Development
Shein-Chung Chow

Randomization, Masking, and Allocation Concealment
Vance W. Berger

Randomized Clinical Trials of Nonpharmacological Treatments
Isabelle Boutron, Philippe Ravaud, and David Moher

Randomized Phase II Cancer Clinical Trials
Sin-Ho Jung

Repeated Measures Design with Generalized Linear Mixed Models for Randomized Controlled Trials
Toshiro Tango

Sample Size Calculations for Clustered and Longitudinal Outcomes in Clinical Research
Chul Ahn, Moonseong Heo, and Song Zhang

Sample Size Calculations in Clinical Research, Third Edition
Shein-Chung Chow, Jun Shao, Hansheng Wang, and Yuliya Lokhnygina

Statistical Analysis of Human Growth and Development
Yin Bun Cheung

Statistical Design and Analysis of Clinical Trials: Principles and Methods
Weichung Joe Shih and Joseph Aisner

Statistical Design and Analysis of Stability Studies
Shein-Chung Chow

Statistical Evaluation of Diagnostic Performance: Topics in ROC Analysis
Kelly H. Zou, Aiyi Liu, Andriy Bandos, Lucila Ohno-Machado, and Howard Rockette

Statistical Methods for Clinical Trials
Mark X. Norleans

Statistical Methods for Drug Safety
Robert D. Gibbons and Anup K. Amatya

Statistical Methods for Healthcare Performance Monitoring
Alex Bottle and Paul Aylin

Statistical Methods for Immunogenicity Assessment
Harry Yang, Jianchun Zhang, Binbing Yu, and Wei Zhao

Statistical Methods in Drug Combination Studies
Wei Zhao and Harry Yang

Statistical Testing Strategies in the Health Sciences
Albert Vexler, Alan D. Hutson, and Xiwei Chen

Statistical Topics in Health Economics and Outcomes Research
Demissie Alemayehu, Joseph C. Cappelleri, Birol Emir, and Kelly H. Zou

Statistics in Drug Research: Methodologies and Recent Developments
Shein-Chung Chow and Jun Shao

Statistics in the Pharmaceutical Industry, Third Edition
Ralph Buncher and Jia-Yeong Tsay

Survival Analysis in Medicine and Genetics
Jialiang Li and Shuangge Ma

Theory of Drug Development
Eric B. Holmgren

Translational Medicine: Strategies and Statistical Methods
Dennis Cosmatos and Shein-Chung Chow

Statistical Topics in Health Economics and Outcomes Research

Edited by
Demissie Alemayehu, PhD
Joseph C. Cappelleri, PhD, MPH, MS
Birol Emir, PhD
Kelly H. Zou, PhD, PStat®

CRC Press
Taylor & Francis Group
Boca Raton London New York

CRC Press is an imprint of the
Taylor & Francis Group, an **informa** business

A CHAPMAN & HALL BOOK

CRC Press
Taylor & Francis Group
6000 Broken Sound Parkway NW, Suite 300
Boca Raton, FL 33487-2742

First issued in paperback 2021

© 2018 by Taylor & Francis Group, LLC
Chapman & Hall is an imprint of Taylor & Francis Group, an Informa business

No claim to original U.S. Government works

ISBN-13: 978-1-03-209604-9 (pbk)
ISBN-13: 978-1-4987-8187-9 (hbk)

Library of Congress Cataloging-in-Publication Data

Names: Alemayehu, Demissie, editor.
Title: Statistical topics in health economics and outcomes research / edited by Demissie Alemayehu, Joseph C. Cappelleri, Birol Emir, Kelly H. Zou.
Description: Boca Raton, Florida : CRC Press, [2018] | Includes bibliographical references and index.
Identifiers: LCCN 2017032464| ISBN 9781498781879 (hardback : acid-free paper) | ISBN 9781498781886 (e-book)
Subjects: LCSH: Medical economics--Statistical methods. | Medical economics--Data processing. | Clinical trials.
Classification: LCC RA410 .S795 2018 | DDC 338.4/73621--dc23
LC record available at https://lccn.loc.gov/2017032464

Visit the Taylor & Francis Web site at
http://www.taylorandfrancis.com

and the CRC Press Web site at
http://www.crcpress.com

Table of Contents

Preface ix
Acknowledgment xiii
About the Editors xv
Authors' Disclosures xvii

1. Data Sources for Health Economics and
   Outcomes Research 1
   Kelly H. Zou, Christine L. Baker, Joseph C. Cappelleri,
   and Richard B. Chambers

2. Patient-Reported Outcomes: Development and Validation 15
   Joseph C. Cappelleri, Andrew G. Bushmakin, and Jose Ma. J. Alvir

3. Observational Data Analysis 47
   Demissie Alemayehu, Marc Berger, Vitalii Doban, and Jack Mardekian

4. Predictive Modeling in HEOR 69
   Birol Emir, David C. Gruben, Helen T. Bhattacharyya,
   Arlene L. Reisman, and Javier Cabrera

5. Methodological Issues in Health Economic Analysis 85
   Demissie Alemayehu, Thomas Mathew, and Richard J. Willke

6. Analysis of Aggregate Data 123
   Demissie Alemayehu, Andrew G. Bushmakin, and Joseph C. Cappelleri

7. Health Economics and Outcomes Research in
   Precision Medicine 151
   Demissie Alemayehu, Joseph C. Cappelleri, Birol Emir,
   and Josephine Sollano

8. Best Practices for Conducting and Reporting
   Health Economics and Outcomes Research 177
   Kelly H. Zou, Joseph C. Cappelleri, Christine L. Baker,
   and Eric C. Yan

Index 185

Preface

With the ever-rising costs associated with health care, evidence generation through health economics and outcomes research (HEOR) plays an increasingly important role in decision-making regarding the allocation of scarce resources. HEOR aims to address the comparative effectiveness of alternative interventions and their associated costs using data from diverse sources and rigorous analytical methods.

While there is a great deal of literature on HEOR, there appears to be a need for a volume that presents a coherent and unified review of the major issues that arise in application, especially from a statistical perspective. Accordingly, this monograph is intended to fill a literature gap in this important area by way of giving a general overview on some of the key analytical issues. As such, this monograph is intended for researchers in the health care industry, including those in the pharmaceutical industry, academia, and government, who have an interest in HEOR. This volume can also be used as a resource by both statisticians and nonstatisticians alike, including epidemiologists, outcomes researchers, and health economists, as well as health care policy- and decision-makers.

This book consists of stand-alone chapters, with each chapter dedicated to a specific topic in HEOR. In covering topics, we made a conscious effort to provide a survey of the relevant literature, and to highlight emerging and current trends and guidelines for best practices, when the latter were available. Some of the chapters provide additional information on pertinent software to accomplish the associated analyses.

Chapter 1 provides a discussion of evidence generation in HEOR, with an emphasis on the relative strengths of data obtained from alternative sources, including randomized control trials, pragmatic trials, and observational studies. Recent developments are noted.

Chapter 2 canvasses a thorough exposition of pertinent aspects of scale development and validation for patient-reported outcomes (PROs). Topics covered include descriptions and examples of content validity, construct validity, and criterion validity. Also covered are exploratory factor analysis and confirmatory factor analysis, two model-based approaches commonly used for validity assessment. Person-item maps are featured as a way to visually and numerically examine the validity of a PRO measure. Furthermore, reliability is discussed in terms of reproducibility of measurement.

The focus of Chapter 3 is the role of observational studies in evidence-based medicine. This chapter highlights steps that need to be taken to maximize their evidentiary value in promoting public health and advancing research in medical science. The issue of confounding in causal inference is discussed, along with design and analytical considerations concerning real-world data. Selected examples of best practices are provided, based on a survey of the available literature on analysis and reporting of observational studies.

Chapter 4 provides a high-level overview of predictive modeling approaches, including linear and nonlinear models, as well as tree-based methods. Applications in HEOR are illustrated, and available software packages are discussed.

The theme of Chapter 5 is cost-effectiveness analysis (CEA), which plays a critical role in health care decision-making. Methodological issues associated with CEA are discussed, and a review of alternative approaches is provided. The chapter also describes the incorporation of evidence through indirect comparisons, as well as data from noninterventional studies. Special reference is made to the use of decision trees and Markov models.

In Chapter 6, statistical issues that arise when synthesizing information from multiple studies are addressed, with reference to both traditional meta-analysis and the emerging area of network meta-analysis. Formal expressions of the underlying models are provided, with a thorough discussion of the relevant assumptions and measures that need to be taken to mitigate the impacts of deviations from those assumptions. In addition, a brief review of the recent literature on best practices for the conduct and reporting of such studies is provided. Also featured is an illustration of random effects meta-analysis using simulated data.

Chapter 7 presents challenges and opportunities of precision medicine in the context of HEOR. Here, it is noted that effective assessment on the cost-benefit of personalized medicines requires addressing fundamental regulatory and methodological issues, including the use of state-of-the-science analytical techniques, the improvement of HEOR data assessment pathways, and the understanding of recent advances in genomic biomarker development. Notably, analytical issues and approaches pertaining to subgroup analysis, as well as genomic biomarker development, are summarized. The role of PRO measures in personalized medicines is discussed. In addition, reference is made to regulatory, market access, and other aspects of personalized medicine. Illustrative examples are provided, based on a review of the recent literature.

Finally, Chapter 8 features some best practices and guidelines for conducting and reporting data from HEOR. Several guidance resources are highlighted, including those from the International Society for Pharmacoeconomics and Outcomes Research (ISPOR), and other professional and governmental bodies.

Given the breadth of the topics in HEOR, it is understood that this volume may not be viewed as a comprehensive reference for all the issues that need to be tackled in practice. Nonetheless, it is hoped that this monograph can still serve a useful purpose in raising awareness about critical issues and in providing guidance to ensure the credibility and strength of HEOR data used in health care decision-making.

D.A., J.C.C., B.E. & K.H.Z., Co-editors

Acknowledgment

The authors are grateful to colleagues for reviewing this document and providing constructive comments. Special thanks go to Linda S. Deal for critiquing the chapter on PROs and to an anonymous reviewer for constructive, helpful comments that improved the quality of several chapters.

About the Editors

Demissie Alemayehu, PhD, is Vice President and Head of the Statistical Research and Data Science Center at Pfizer Inc. He earned his PhD in Statistics from the University of California at Berkeley. He is a Fellow of the American Statistical Association, has published widely, and has served on the editorial boards of major journals, including the *Journal of the American Statistical Association* and the *Journal of Nonparametric Statistics*. Additionally, he has been on the faculties of both Columbia University and Western Michigan University. He has co-authored *Patient-Reported Outcomes: Measurement, Implementation and Interpretation*, published by Chapman & Hall/CRC Press.

Joseph C. Cappelleri, PhD, MPH, MS is Executive Director at the Statistical Research and Data Science Center at Pfizer Inc. He earned his MS in Statistics from the City University of New York, PhD in Psychometrics from Cornell University, and MPH in Epidemiology from Harvard University. As an adjunct professor, he has served on the faculties of Brown University, the University of Connecticut, and Tufts Medical Center. He has delivered numerous conference presentations and has published extensively on clinical and methodological topics, including regression-discontinuity designs, meta-analyses, and health measurement scales. He is lead author of the monograph *Patient-Reported Outcomes: Measurement, Implementation and Interpretation*. He is a Fellow of the American Statistical Association.

Birol Emir, PhD, is Senior Director and Statistics Lead at the Statistical Research and Data Science Center at Pfizer Inc. In addition, he is an Adjunct Professor of Statistics at Columbia University in New York and an External PhD Committee Member at the Graduate School of Arts and Sciences at Rutgers, The State University of New Jersey. Recently, his primary focuses have been on big data, predictive modeling, and genomics data analysis. He has numerous publications in refereed journals, and authored a book chapter in *A Picture Is Worth a Thousand Tables: Graphics in Life Sciences*. He has taught several short courses and has given invited presentations.

Kelly H. Zou, PhD, PStat®, is Senior Director and Analytic Science Lead at Pfizer Inc. She is a Fellow of the American Statistical Association and an Accredited Professional Statistician. She has published extensively on clinical and methodological topics. She has served on the editorial board of *Significance*, as an Associate Editor for *Statistics in Medicine* and *Radiology*, and as a Deputy Editor for *Academic Radiology*. She was Associate Professor of Radiology, Director of Biostatistics, and Lecturer of Health Care

Policy at Harvard Medical School. She was Associate Director of Rates at Barclays Capital. She has co-authored *Statistical Evaluation of Diagnostic Performance: Topics in ROC Analysis* and *Patient-Reported Outcomes: Measurement, Implementation and Interpretation*, both published by Chapman and Hall/CRC. She authored a book chapter in *Leadership and Women in Statistics* by the same publisher. She was the theme editor on a statistics book titled *Mathematical and Statistical Methods for Diagnoses and Therapies*.

Authors' Disclosures

Demissie Alemayehu, Jose Ma. J. Alvir, Christine L. Baker, Marc Berger, Helen T. Bhattacharyya, Andrew G. Bushmakin, Joseph C. Cappelleri, Richard B. Chambers, Vitalii Doban, Birol Emir, David C. Gruben, Jack Mardekian, Arlene L. Reisman, Eric C. Yan, and Kelly H. Zou are employees of Pfizer Inc. Josephine Sollano is a former employee of Pfizer Inc. This book was prepared by the authors in their personal capacity. The views and opinions expressed in this book are the authors' own, and do not necessarily reflect those of Pfizer Inc.

Additional authors include Javier Cabrera of Rutgers, The State University of New Jersey; Thomas Mathew of the University of Maryland Baltimore County; and Richard J. Willke of the International Society for Pharmacoeconomics and Outcomes Research (ISPOR).

1

Data Sources for Health Economics and Outcomes Research

Kelly H. Zou, Christine L. Baker, Joseph C. Cappelleri, and Richard B. Chambers

CONTENTS

1.1 Introduction 1
1.2 Data Sources and Evidence Hierarchy 2
1.3 Randomized Controlled Trials 3
1.4 Observational Studies 5
1.5 Pragmatic Trials 7
1.6 Patient-Reported Outcomes 8
1.7 Systematic Reviews and Meta-Analyses 9
1.8 Concluding Remarks 10
References 11

1.1 Introduction

The health care industry and regulatory agencies rely on data from various sources to assess and enhance the effectiveness and efficiency of health care systems. In addition to randomized controlled trials (RCTs), alternative data sources such as pragmatic trials and observational studies may help in evaluating patients' diagnostic and prognostic outcomes (Ford and Norrie, 2016). In particular, observational data are increasingly gaining usefulness in the development of policies to improve patient outcomes, and in health technology assessments (Alemayehu and Berger, 2016; Berger and Doban, 2014; Groves et al., 2013; Holtorf et al., 2012; Vandenbroucke et al., 2007; Zikopoulos et al., 2012). However, in view of the inherent limitations, it is important to appropriately apply and systematically evaluate the widespread use of real-world evidence, particularly in the drug approval process.

As a consequence of the digital revolution, medical evidence generation is evolving, with many possible data sources, for example, digital data from the government and private organizations (e.g., health care organizations, payers, providers, and patients) (Califf et al., 2016). A list of different types of research data, with their advantages and disadvantages, may be found in the Himmelfarb Health Sciences Library (2017), which is maintained by the George Washington University.

In this chapter, we provide a brief introduction of some common data sources encountered and analyzed in health economics and outcomes research (HEOR) studies, which include randomized controlled trials (RCTs), pragmatic trials, observational studies, and systematic reviews.

1.2 Data Sources and Evidence Hierarchy

Murad et al. (2016) and Ho et al. (2008) provide the hierarchy or strength of evidence generated from different data sources. According to this hierarchy, depicted in the evidence pyramid in Figure 1.1, a systematic review/meta-analysis of randomized controlled trials (RCTs) and individual RCTs provides the strongest level of evidence, followed by cohorts, case-control studies, cross-sectional studies, and, finally, case series. In particular, prospective cohort studies are generally favored over retrospective cohort studies with regards to strength of evidence.

FIGURE 1.1
Evidence pyramid. (Modified from Dartmouth Biomedical Libraries. *Evidence-based medicine (EBM) resources.* http://www.dartmouth.edu/~biomed/resources.htmld/guides/ebm_resources.shtml, 2017.)

The Enhancing the QUAlity and Transparency Of health Research (EQUATOR) Network (2017) is an international initiative that seeks to improve the reliability and value of published health research literature by promoting transparent and accurate reporting and a wider use of robust reporting guidelines. It is the first coordinated attempt to tackle the problems of inadequate reporting systematically and on a global scale; it advances the work done by individual groups over the last 15 years. The EQUATOR's (2017) website includes guidelines for the following main study types: randomized trials, observational studies, systematic reviews, case reports, qualitative research, diagnostic/prognostic studies, quality improvement studies, economic evaluation, animal/preclinical studies, study protocols, and clinical practice guidelines.

1.3 Randomized Controlled Trials

The RCT was first used in 1948, when the British Medical Research Council (MRC) evaluated streptomycin for treating tuberculosis (Bothwell and Podolsky, 2016; Holtorf, 2012; Sibbald and Roland, 1998). A well-conducted RCT design is generally considered to be the gold standard in terms of providing evidence, because causality can be inferred due to the design's comparisons of randomized groups that are balanced on known and unknown baseline characteristics (Bothwell and Podolsky, 2016). In addition, RCT studies are conducted under controlled conditions with well-defined inclusion and exclusion criteria. Hence, RCTs are the strongest in terms of internal validity and for identifying causation (i.e., making inferences relating to the study population).

Frequently, a placebo group serves as the control group; however, the use of an active control, such as standard of care, is becoming more common. The expected difference on the primary outcome of interest between the interventional group(s) and the control group is the central objective. Typical endpoints include the mean change from baseline, the percent change, and the median time to an event, such as disease recurrence.

A double-blind design is often used in RCTs of pharmaceutical interventions, where assignments into the intervention and the control groups are not known in advance by both investigators and patients. This methodological framework minimizes possible bias that might result from awareness of the treatment group.

According to the Food and Drug Administration (FDA, 2017), the numbers of volunteers across several phases of RCTs are as follows: Phase 1: 20 to 100; Phase 2: several hundred; Phase 3: 300 to 3000; and Phase 4: several thousand. Further details about these phases are also described.

In addition, according to the National Library of Medicine's (NLM, 2017) clinical trial registration site, ClinicalTrials.gov, there are five phases of clinical trials involved in drug development. Phase 0 contains exploratory studies involving very limited human exposure to the drug, with no therapeutic or diagnostic goals (e.g., screening studies, micro-dose studies). Phase 1 involves studies that are usually conducted with healthy volunteers, and emphasize safety. The goals of Phase I studies are to find out what the drug's most frequent and serious adverse events are and, often, how the drug is metabolized and excreted. Phase 2 includes studies that gather preliminary data on efficacy (whether the drug works in people who have a particular disease or condition under a certain set of circumstances). For example, participants receiving the drug may be compared with similar participants receiving a different treatment, usually an inactive substance, called a placebo, or a standard therapy. Drug safety also continues to be evaluated in Phase 2, and short-term adverse events are studied.

Phase 3 includes confirmatory studies for the purpose of regulatory approval and gather more information about the efficacy and safety by studying targeted populations, with possibly different dosages and drug combinations. These studies are typically much larger in size than the Phase 2 studies, and are often multinational. Phase 4 contains studies that occur after the Food and Drug Administration (FDA) has approved a drug for marketing. These studies involve a postmarket investigation to sponsored studies required of or agreed to by the study sponsor for the purpose of gathering additional information about a drug's safety, efficacy, or optimal use scenarios, including its use in subgroups of patients.

The numbers of volunteers were as follows: Phase 1: 20 to 100; Phase 2: several hundred; Phase 3: 300 to 3000; and Phase 4: several thousand (https://www.fda.gov/ForPatients/Approvals/Drugs/ucm405622.htm#Clinical_Research_Phase_Studies).

Over the last few decades, the use of a particular type of RCT—the multicenter clinical trial—has become quite popular. As a result of a potentially long enrollment period, trial enrollment may benefit from simultaneous patient recruitment from multiple sites, which may be within a country or region. Pharmaceutical and biotechnology companies and parts of the US National Institutes of Health (NIH), such as the National Cancer Institute, have been among the sponsors of multicenter clinical trials. Such large and complex studies require sophisticated data management, analysis, and interpretation.

ClinicalTrials.gov of the US NLM (2017) is a registry and results database of publicly and privately supported clinical studies of human participants that have or are being conducted around the world. It allows the public to learn more about clinical studies through information on its website and provides background information on relevant history, policies, and laws. In April 2017, this website listed approximately 240,893 studies with locations in all 50 US states and in 197 countries. According to the displayed enrollment status, the locations of recruiting studies are as follows: non-US-only (56%), US-only (38%), and both US and non-US (5%). Thus, most of the registered studies are conducted outside of the United States.

It is noted that RCTs are compromised with respect to external validity (i.e., making inferences outside of the study population or testing conditions), since the conditions under which they are conducted do not necessarily reflect the real world, with its inherent complexity and heterogeneity. Accordingly, data from nonrandomized studies may need to be used to complement RCTs or to fill the evidentiary gap created by the unavailability of RCT data.

1.4 Observational Studies

Section 3022 in the 21st Century Cures Act of the United States Congress (United States Congress, 2016) defines the term "real-world evidence" (RWE) to mean "data regarding the usage, or the potential benefits or risks, of a drug derived from sources other than randomized clinical trials." Accordingly, noninterventional (observational) studies do not involve randomizations, but can provide real-world data (RWD) to generate RWE. Such studies are selected due to their ease of implementation, cost considerations, and generalizability from broad experiences (Garrison, 2007). Sherman et al. (2016) indicate RWE includes "information on health care that is derived from multiple sources outside typical clinical research settings, including electronic health records (EHRs), claims and billing data, product and disease registries, and data gathered through personal devices and health applications."

RWD generally represent the various complex treatment choices, switches between the treatments, the strengths (dose levels), and the days of supply (pill counts) that are often found in actual clinical practice. Furthermore, adding patients' demographic characteristics, comorbidities, concomitant treatments, and switches between therapies provides a real-world understanding of the magnitude and variability of the treatment's effect in different sets of circumstances.

Observational studies involve existing databases, with standardized methodologies employed depending on the objective of the question being evaluated. Use of this methodological framework can be both practical and convenient and, in addition, prospective or retrospective. Cohort studies, cross-sectional studies, and case-control studies are among the different types of study designs included within the umbrella of observational studies (Mann, 2003). Retrospective cohort databases can help patients, health care providers, and payers understand the epidemiology of a disease or an unmet medical need. They inform in several important areas, for example, in precision medicine for drug discovery and development, by examining baseline patient characteristics and comorbid conditions; in quality improvement or efficiency improvement efforts and in health technology assessments or decisions regarding access to and about the pricing of new therapies; and in bedside shared decision-making between patients and their providers.

Retrospective cohort studies can also facilitate access to the incidence or prevalence of adverse events associated with marketed medications to inform regulatory labeling (Garrison et al., 2007).

With the increasing availability of big data, structured and unstructured data, digital media, images, records, and free texts, there is an abundance of databases for designing and implementing observational studies. Thus, improvements in the storage, archiving, and sharing of information may make observational studies increasingly more attractive. Data mining, machine learning, and predictive modeling algorithms, as described in subsequent chapters of this book, also contribute to the increasing popularity of these approaches.

Unlike RCT data, however, observational data can be collected in routine clinical practice or via administrative claims. Therefore, these data are collected without being generated based on investigators' scientific hypotheses in mind. Although these data are conveniently available, there may likely be sampling biases, missing or incomplete data, and data entry errors that need to be addressed. In order to guard and protect individual patients' privacy, the de-identification of the datasets should be undertaken by removing sensitive identifiable information across patients.

According to a task force created by the International Society for Pharmacoeconomics and Outcomes Research (ISPOR), Garrison et al. (2007) defined RWD to mean "data used for decision-making that are not collected in conventional RCTs." To characterize RWD, these authors suggested three approaches based on (1) the types of outcomes (clinical, economic, and patient-reported); (2) the hierarchies of evidence (RCT, observational data, and so on); (3) the data sources used. These data sources include supplements to traditional registration RCTs; large simple trials that "are primarily Phase IV studies that are embedded in the delivery of care, make use of EHRs, demand little extra effort of physicians and patients, and can be conducted for a relatively modest sum"; patient registries; administrative data; health surveys; EHRs including medical chart reviews (Roehr, 2013).

For conducting research based on observational data, a protocol with a prespecified statistical analysis plan ideally should be developed. For example, the Agency for Healthcare Quality and Research (AHRQ, 2013) has crafted and recommended comprehensive protocol design elements.

It is also important to develop access to RWD by building the appropriate infrastructure. Query tools for rapid and in-depth data analyses are at the forefront of RWD collaboration. Data sharing is another efficient way to streamline the lengthy and costly development and clinical trial processes. For example, the RWE and pragmatic trials may be used to supplement the results obtained from a costly RCT alone.

RWD can provide opportunities for effective collaborations and partnerships among academia, industry, and government to unlock the value of big data in health care. However, in the opinion of the authors of this chapter, there is a list of potential challenges to overcome in order to build a strong infrastructure and to adequately meet talent requirements, as well as quality and standard variables through connected datasets. Well-defined and scientifically sound research questions have been proposed regarding disease burden, health and quality of life outcomes, utilization methods, and costs on both the population level and the individual level (Willke and Mullins, 2011).

Several methodological challenges exist. Among these challenges are (1) the maintenance of privacy and security regarding data access and the data governance model; (2) the linkage of different sources including novel sources, biobanks and genomics, social media and credit information, sensors/wearables, soluble problems, and the growing number of data aggregators; and (3) addressing other aspects of analytics, imputing causation versus correlation, and considering emerging approaches to increasingly complex problems.

RWD and the vast datasets being developed and shared can help to shorten clinical trial times and decrease costs related to bringing a therapy to the market. The Collaboratory Distributed Research Network (DRN) of the US NIH (2016) enables investigators to collaborate with each other in the use of electronic health data, while also safeguarding protected health information and proprietary data. It supports both single- and multisite research programs. Its querying capabilities reduce the need to share confidential or proprietary data by enabling authorized researchers to send queries to collaborators holding data such as data partners. In some cases, queries can take the form of computer programs that a data partner can execute on a preexisting dataset. The data partner can return the query result, typically aggregated data (e.g., count data) rather than the raw data itself. Remote querying reduces legal, regulatory, privacy, proprietary, and technical barriers associated with sharing data for research. Example data sharing models include the Mini-Sentinel (2016), AMCP Task Force on Biosimilar Collective Intelligence Systems et al. (2015), Observational Medical Outcomes Partnership (OMOP, 2016), and the Biologics and Biosimilars Collective Intelligence Consortium (BBCIC, 2017).

1.5 Pragmatic Trials

In contrast to an RCT, which consists of controlled experimental conditions, pragmatic trials are randomized and minimally controlled studies intended to "measure effectiveness—the benefit the treatment produces in routine clinical practice" (Patsopoulos, 2011; Roland, 1998). Pragmatic trials can be considered a special type of observation study. This type of design extends the testing conditions to the real-world setting, which has greater complexity, rather than only to the limited and controlled conditions inherent in the RCT. Thus, there are considerable opportunities to conduct pragmatic trials with an observational data component (Ford and Norrie, 2016).

As highlighted previously in this chapter, explanatory trials, such as RCTs, aim to confirm a prespecified hypothesis in a given target population. In contrast, however, pragmatic trials "inform a clinical or policy decision by providing evidence for adoption of the intervention into real world clinical practice." (Ford and Norrie, 2016; Roland, 1998; Schwartz and Lellouch, 1967).

The main advantage of pragmatic studies is that they address practical questions about the risks, benefits, and costs of an intervention versus the usual care in routine clinical practice. Specific RWD in patient populations are useful to health care providers, patients, payers, and other decision-makers (Mentz et al., 2016; Patsopoulos, 2011; Whicher et al., 2015). Such data provide evidence for expected outcomes in a typical patient population with typical adherence.

While pragmatic studies are often randomized, they are otherwise less controlled and more realistic than the standard RCT. Sherman et al. (2016) stated that "in addition to its application in interventional studies, real world evidence is also valuable in observational settings, where it is used to generate hypotheses for prospective trials, assess the generalizability of findings from interventional trials (including RCTs), conduct safety surveillance of medical products, examine changes in patterns of therapeutic use, and measure and implement quality in health care delivery." Once patients are assigned to the treatment group, pragmatic studies have fewer controlled conditions (e.g., established clinic visits or telephone contacts as would occur in an RCT) prior to the evaluation of the study outcome. Additional limitations of pragmatic studies are that there may be an increased amount of missing data, biases, and other less stringent enrollment issues as compared with RCTs (Mentz et al., 2016). Generally, regulators may have some reservation in using this design to make decisions on efficacy and safety because of its lower evidence tier than that of an RCT. Therefore, it is important to clearly explain the pros and cons of the pragmatic study design when communicating with regulatory bodies and agencies (Anderson et al., 2015; Maclure, 2009). There are two useful tools to determine how pragmatic a particular study is: the Pragmatic–Explanatory Continuum Indicator Summary (PRECIS) and PRECIS-2. Further details on these indicators can be found in Ford and Norrie (2016) and Sherman et al. (2016).

1.6 Patient-Reported Outcomes

A patient-reported outcome (PRO) is any report on the status of a patient's health condition that comes directly from the patient, without interpretation of the patient's response by a clinician or anyone else (FDA, 2009). It can be measured in an RCT, or can be derived from an observational study.

Thus, a PRO can be part of several hierarchies in the evidence pyramid. As an umbrella term, PROs include a whole host of subjective outcomes. A few specific examples include pain, fatigue, depression, aspects of well-being (e.g., physical, functional, psychological), treatment satisfaction, health-related quality of life, and physical symptoms (e.g., nausea and vomiting).

PROs are often relevant in studying a variety of conditions, including pain, erectile dysfunction, fatigue, migraine, mental functioning, physical functioning, and depression, which cannot be assessed adequately without a patient's evaluation and whose key questions require a patient's input on the impact of a disease or its treatment. Data generated by a PRO instrument can provide a statement of a treatment benefit from the patient perspective (Cappelleri et al., 2013; de Vet et al., 2011; Fayers and Machin, 2016; Streiner et al., 2015). For a treatment benefit to be meaningful, though, the PRO under consideration must be validated, meaning there should be evidence that it effectively measures the particular concept under study; that is, it measures what it is intended to measure, and does so reliably.

1.7 Systematic Reviews and Meta-Analyses

Meta-analysis refers to the practice of applying statistical methods to combine and quantify the outcomes of a series of studies in a single pooled analysis. It is part of a quantitative systematic overview. The Cochrane Consumer Network (2017) states the following: "A systematic review summarizes the results of available carefully designed health care studies (controlled trials) and provides a high level of evidence on the effectiveness of health care interventions. Judgments may be made about the evidence and inform recommendations for health care to summarize the results of available carefully designed health care studies (controlled trials) and provides a high level of evidence on the effectiveness of health care interventions. Judgments may be made about the evidence and inform recommendations for health care." Additionally, it employs specific analytic methods for combining pertinent quantitative results from multiple selected studies to develop an overall estimate with its accompanying precision. The Cochrane Library (2017) provides a set of training items about the foundational concepts associated with both systematic review and meta-analysis.

Meta-analysis is used for the following purposes: (1) to establish statistical significance with studies that have conflicting results; (2) to develop a more correct or refined estimate of effect magnitude; (3) to provide a more comprehensive assessment of harms, safety data, and benefits; and (4) to examine subgroups with a larger sample size than any one study (Uman, 2011).

Conclusions from well-conducted and high-quality meta-analyses result in stronger evidence than those from a single study because of the increased numbers of subjects, greater ability to discern heterogeneity of results among different types of patients and studies, and accumulated effects and results.

Because it does not use statistical methods for pooling results, and tends to summarize more in qualitative (narrative) rather than in quantitative terms, the narrative review cannot be regarded as a meta-analysis. There is a distinction that needs to be made between exploratory and confirmatory use of meta-analyses. Most published meta-analyses are performed retrospectively, after the data and results are available. Unless the meta-analysis is planned in advance (as a prospective meta-analysis), it is unlikely that regulatory authorities will accept it as a definitive proof of effect. There are a number of uses to which meta-analysis can be put in an exploratory way. Meta-analyses are being increasingly applied to generate hypotheses regarding safety outcomes (adverse events), where there are special challenges beyond those found for efficacy outcomes (Bennetts et al., 2017).

In choosing a meta-analytic framework with a fixed effects model or a random effects model, it is important to realize that each model addresses a different research question. If the research question is concerned with an overall treatment effect in the existing studies, and there is evidence that there is a common treatment effect across studies, only the variability within a study is required to answer whether the size of the observed effect is consistent with chance or not. From this perspective, meta-analysis is not concerned with making an inference to a larger set of studies, and the use of a fixed effects model would be appropriate. If one wants to estimate the treatment effect that would be observed in a future study, while allowing for studies to have their own treatment effects distributed around a central value, then the heterogeneity of the treatment effect across studies should be accounted for with a random effects model, which incorporates not only within-study variability of the treatment effect, but also between-study variability of the treatment effect.

1.8 Concluding Remarks

This chapter provides a broad account on how to generate accurate, representative, and reliable evidence. In doing so, this chapter highlights the various types of data that this book will examine and illustrate in subsequent chapters. Methodologies must be carefully selected and findings must be appropriately interpreted to provide strong support for claims in publications, approved medical product labeling, and market access.

References

Agency for Healthcare Research and Quality (AHRQ). 2013. *Developing a protocol for observational comparative effectiveness research: A user's guide.* http://www.effectivehealthcare.ahrq.gov/ehc/products/440/1166/User-Guide-to-Observational-CER-1-10-13.pdf (accessed May 11, 2017).

Alemayehu, D. and M. Berger. 2016. Big data: Transforming drug development and health policy decision making. *Health Serv Outcomes Res Methodol* 16:92–102.

AMCP Task Force on Biosimilar Collective Intelligence Systems, Baldziki, M., Brown, J. et al. 2015. Utilizing data consortia to monitor safety and effectiveness of biosimilars and their innovator products. *J Manag Care Spec Pharm* 21:23–34.

Anderson, M. L., Griffin, J., Goldkind, S. F. et al. 2015. The Food and Drug Administration and pragmatic clinical trials of marketed medical products. *Clin Trials* 12:511–519.

Bennetts, M., Whalen, E., Ahadieh, S. et al. 2017. An appraisal of meta-analysis guidelines: How do they relate to safety outcomes? *Res Synth Methods* 8:64–78.

Berger, M.L. and V. Doban. 2014. Big data, advanced analytics and the future of comparative effectiveness research. *J Comp Eff Res* 3:167–176.

Biologics and Biosimilars Collective Intelligence Consortium (BBCIC). 2017. http://bbcic.org (accessed May 11, 2017).

Bothwell, L.E. and S.H. Podolsky. 2016. The emergence of the randomized, controlled trial. *N Engl J Med* 375:501–504.

Califf, R.M., Robb, M.A., Bindman, A.B. et al. 2016. Transforming evidence generation to support health and health care decisions. *N Engl J Med* 375:2395–2400.

Cappelleri, J.C., Zou, K.H., Bushmakin, A.G. et al. 2013. *Patient-reported outcomes: Measurement, implementation and interpretation.* Boca Raton, FL: CRC Press/Taylor & Francis.

ClinicalTrials.gov. 2017. *Advanced search field definitions.* https://clinicaltrials.gov/ct2/help/how-find/advanced/field-defs (accessed May 11, 2017).

Cochrane Consumer Network. 2017. *What is a systematic review?* http://consumers.cochrane.org/what-systematic-review (accessed May 11, 2017).

Cochrane Library. 2017. *Introduction to systematic reviews: Online learning module, Cochrane Training.* http://training.cochrane.org/resource/introduction-systematic-reviews-online-learning-module (accessed May 11, 2017).

Dartmouth Biomedical Libraries. 2017. *Evidence-based medicine (EBM) resources.* http://www.dartmouth.edu/~biomed/resources.htmld/guides/ebm_resources.shtml (accessed May 11, 2017).

de Vet, H.C.W., Terwee, C.B., Mokkink, L.B. et al. 2011. *Measurement in Medicine: A Practical Guide.* New York, NY: Cambridge University Press.

EQUATOR Network. 2017. *Enhancing the QUAlity and Transparency Of health Research.* http://www.equator-network.org (accessed May 11, 2017).

Fayers, P.M. and D. Machin. 2016. *Quality of Life: The Assessment, Analysis and Reporting of Patient-Reported Outcomes.* 3rd ed. Chichester, UK; John Wiley & Sons, Ltd.

Ford, I. and J. Norrie. 2016. Pragmatic trials. *N Engl J Med* 375:454–463.

Food and Drug Administration (FDA). 2009. Guidance for industry on patient-reported outcome measures: Use in medical product development to support labeling claims. *Federal Register* 74(235):65132–65133.

Food and Drug Administration (FDA). 2017. The Drug Development Process: Step 3: Clinical Research. https://www.fda.gov/ForPatients/Approvals/Drugs/ucm405622. htm#Clinical_Research_Phase_Studies (assessed May 11, 2017).

Garrison Jr., L.P., Neumann, P.J., Erickson, P. et al. 2007. Using real-world data for coverage and payment decisions: the ISPOR Real-World Data Task Force report. *Value Health* 10:326–335.

Groves, P., Kayyali, B., Knott, D. et al. 2013. *The 'big data' revolution in healthcare.* McKinsey & Company, Center for US Health System Reform Business Technology Office. http://www.mckinsey.com/industries/healthcare-systems-and-services/our-insights/the-big-data-revolution-in-us-health-care (accessed May 11, 2017).

Himmelfarb Health Sciences Library. 2017. *Welcome to study design 101.* The George Washington University. https://himmelfarb.gwu.edu/tutorials/studydesign101/index.html (accessed May 11, 2017).

Ho, P.M., Peterson, P.N. and Masoudi, F.A. 2008. Evaluating the evidence: Is there a rigid hierarchy? *Circulation* 118:1675–1684.

Holtorf, A.P., Brixner, D., Bellows, B. et al. 2012. Current and future use of HEOR data in healthcare decision-making in the United States and in emerging markets. *Am Health Drug Benefits* 5:428–438.

Maclure, M. 2009. Explaining pragmatic trials to pragmatic policy-makers. *CMAJ* 180:1001–1003.

Mann, C.J. 2003. Observational research methods. Research design II: cohort, cross sectional, and case-control studies. *Emerg Med J* 20:54–60.

Mentz, R.J., Hernandez, A.F., Berdan, L.G. et al. 2016. Good clinical practice guidance and pragmatic clinical trials: Balancing the best of both worlds. *Circulation* 133:872–880.

Mini-Sentinel. 2017. http://mini-sentinel.org/data_activities/distributed_db_and_data/default.aspx (accessed May 11, 2017).

Murad, M.H., Asi, N., Alsawas, M. et al. 2016. New evidence pyramid. *Evid Based Med* 21:125–127.

National Institutes of Health (NIH). 2016. *NIH collaboratory distributed research network (DRN).* https://www.nihcollaboratory.org/Pages/distributed-research-network.aspx (accessed May 11, 2017).

National Library of Medicine (NLM). 2017. *ClinicalTrials.gov.* https://clinicaltrials.gov (accessed May 11, 2017).

Observational Medical Outcomes Partnership (OMOP). 2016. http://omop.org (accessed May 11, 2017).

Patsopoulos, N.A. 2011. A pragmatic view on pragmatic trials. *Dialogues Clin Neurosci* 13:217–224.

Roehr, B. 2013. The appeal of large simple trials. *BMJ* 346:f1317.

Roland, M. 1998. Understand controlled trials: What are pragmatic trials? *BMJ* 316:285.

Schwartz, D. and J. Lellouch. 1967. Explanatory and pragmatic attitudes in therapeutical trials. *J Chronic Dis* 20:637–648.

Sherman, R.E., Anderson, S.A., Dal Pan, G.J. et al. 2016. Real-world evidence -what is it and what can it tell us? *N Engl J Med* 375:2293–2297.

Sibbald, B. and M. Roland. 1998. Understanding controlled trials. Why are randomised controlled trials important? *BMJ* 316:201.

Streiner, D.L., Norman, G.R. and J. Cairney. 2015. *Health Measurement Scales: A Practical Guide to Their Development and Use.* 5th ed. New York, NY: Oxford University Press.

Uman, L.S. 2011. Systematic reviews and meta-analyses. *J Can Acad Child Adolesc Psychiatry* 20:57–59.

United States Congress. 2016. *H.R.34 - 21st Century Cures Act, 114th Congress.* https://www.congress.gov/114/bills/hr34/BILLS-114hr34enr.pdf (accessed May 11, 2017).

Vandenbroucke, J.P., von Elm, E., Altman, D.G. et al. 2007. STROBE initiative. Strengthening the reporting of observational Studies in Epidemiology (STROBE): Explanation and elaboration. *Ann Intern Med* 147: 573–577. (Erratum in: *Ann Intern Med* 148:168.)

Whicher, D. M., Miller, J. E., Dunham, K. M. et al. 2015. Gatekeepers for pragmatic clinical trials. 2015. *Clin Trials* 12:442–448.

Willke, R.J. and C.D. Mullins. 2011. "Ten commandments" for conducting comparative effectiveness research using "real-world data." *J Manag Care Pharm* 17:S10–S15.

Zikopoulos, P.C., Eaton, C., deRoos, D. et al. 2012. *Understanding Big Data: Analytics for Enterprise Class Hadoop and Streaming Data.* New York, NY: McGraw Hill. https://www.ibm.com/developerworks/vn/library/contest/dw-freebooks/Tim_Hieu_Big_Data/Understanding_BigData.PDF (accessed May 11, 2017).

2

Patient-Reported Outcomes: Development and Validation

Joseph C. Cappelleri, Andrew G. Bushmakin, and Jose Ma. J. Alvir

CONTENTS

2.1   Introduction 16
2.2   Content Validity 17
2.3   Construct Validity 19
    2.3.1   Convergent Validity and Divergent Validity 20
    2.3.2   Known-Groups Validity 21
    2.3.3   Criterion Validity 24
2.4   EFA 24
    2.4.1   Role of EFA 25
    2.4.2   EFA Model 25
    2.4.3   Number of Factors 27
    2.4.4   Factor Rotation 28
    2.4.5   Sample Size 28
    2.4.6   Assumptions 29
    2.4.7   Real-Life Application 30
2.5   CFA 31
    2.5.1   EFA versus CFA 31
    2.5.2   Measurement Model 31
    2.5.3   Standard Model versus Nonstandard Model 32
    2.5.4   Depicting the Model 32
    2.5.5   Identifying Residual Terms for Endogenous Variables 33
    2.5.6   Identifying All Parameters to Be Estimated 34
    2.5.7   Assessing Fit between Model and Data 34
    2.5.8   Real-Life Application 35
2.6   Person-Item Maps 36
2.7   Reliability 39
    2.7.1   Repeatability Reliability 39
    2.7.2   Internal Consistency Reliability 41
2.8   Conclusions 42
Acknowledgments 43
References 44

2.1 Introduction

A patient-reported outcome (PRO) is any report on the status of a patient's health condition that comes directly from the patient, without interpretation of the patient's response by a clinician or anyone else (FDA, 2009). As an umbrella term, PRO measures include a whole host of subjective concepts, such as pain; fatigue; depression; aspects of well-being (e.g., physical, functional, psychological); treatment satisfaction; health-related quality of life; and physical symptoms such as nausea and vomiting. PROs are often relevant in studying a variety of conditions—including pain, erectile dysfunction, fatigue, migraine, mental functioning, physical functioning, and depression—that cannot be assessed adequately without a patient's evaluation, and whose key questions require a patient's input on the impact of a disease or its treatment.

Data generated by a PRO measure can provide a statement of treatment benefit from the patient perspective, and can become part of a regulatory label claim for a therapeutic intervention. In addition, PRO measures have merits that go beyond satisfying regulatory requirements for a label claim (Doward et al., 2010). Payers both in the United States and Europe, clinicians, and patients themselves all have an interest in PRO measures that transcend a label claim, and that are based on the best available evidence, for patient-reported symptoms or any other PRO measure. These key stakeholders help to determine the availability, pricing, and value of medicinal products.

PROs have played a central role in comparative effectiveness research (CER), which seeks to explain the differential benefits and harms of alternate methods to prevent, diagnose, treat, and monitor a clinical condition or to improve the delivery of care (Alemayehu et al., 2011). CER encompasses all forms of data, from controlled clinical trials to outside of them (so-called "real-world" data), including clinical practice. Recommendations have been made for incorporating PRO measures in CER as a guide for researchers, clinicians, and policy-makers in general (Ahmed et al., 2012), and in adult oncology in particular (Basch et al., 2012). Emerging changes that may facilitate CER using PROs include the implementation of electronic and personal health records, hospital and population-based registries, and the use of PROs in national monitoring initiatives.

Funding opportunities have expanded for PRO research. For instance, guided by CER principles, the Patient-Centered Outcomes Research Institute (PCORI) has provided a large number of grants of varying monetary amounts to fund research that can help patients (and those who care for them) make better informed decisions about health care. PCORI seeks to fund useful research that is likely to change practice and improve patient outcomes, and focuses on sharing the resulting information with the public. Moreover, PCORI works to influence research funded by others so that it will become more useful to patients and other health care decision-makers. PROs have therefore become central to patient-centered research and decision-making.

To be useful to patients and other decision-makers (e.g., physicians, regulatory agencies, reimbursement authorities) who are stakeholders in medical care, a PRO measure must undergo a validation process to confirm that it is reliably measuring what it is intended to measure. As assessments of subjective concepts, therefore, PRO measures require evidence of their validity and reliability before they can be used with confidence (Cappelleri et al., 2013; de Vet et al., 2011; Fayers and Machin, 2016; Streiner et al., 2015). Validity assesses the extent to which an instrument measures what it is meant to measure, while reliability assesses how precisely or well the instrument measures what it measures.

The next several sections of this chapter involve the key concepts of validity in the evaluation of a PRO instrument. Section 2.2 covers content validity. Section 2.3 covers construct validity and criterion validity, including their variations. Section 2.4 covers exploratory factor analysis (EFA). Section 2.5 covers confirmatory factor analysis (CFA). Section 2.6 discusses the use of person-item maps as a way to examine validity. Section 2.7 centers on the topic of reliability, which is typically discussed in terms of reproducibility, and is addressed with repeatability reliability and internal consistency reliability. Section 2.8 provides a conclusion.

2.2 Content Validity

There are several forms of validity (Table 2.1) that are discussed in this chapter. In this section, the discussion begins with focusing on content validity.

Instrument development can be an expensive and a time-consuming process. It usually involves a number of considerations: qualitative methods (concept elicitation, item generation, cognitive debriefing, expert panels,

TABLE 2.1

Different Types of Validity

- Content Validity (includes face validity)
- Construct Validity
  - Convergent Validity
  - Divergent (Discriminant) Validity
  - Known-Groups Validity (includes sensitivity and responsiveness)
  - Factor Analysis
- Criterion Validity
  - Concurrent Validity
  - Predictive Validity

qualitative interviews, focus groups); data collection from a sample in the target population of interest; item-reduction psychometric validation; and translation and cultural adaptation. The first and most important step involves the establishment of content validity through qualitative methods—that is, ascertaining whether the measured concepts cover what patients consider the important aspects of the condition and its therapy. The importance of this step cannot be overemphasized.

Rigor in the development of the content of a PRO measure is essential to ensure that the concept of interest is measured accurately, comprehensively, and completely; in order to capture issues of most relevance to the patient; and so as to subscribe to a language that allows for patients to understand and respond without confusion. Items within a questionnaire that have little relevance to the patient population being investigated, or that are poorly written, will lead to measurement error and bias, resulting in ambiguous responses.

Therefore, taking the time to communicate with patients about their symptoms or the impact of a disease or condition on the concept of interest (which the PRO instrument is intended to measure) is very important before embarking on generation of the questions to measure the concept. Qualitative research with patients is essential for establishing content validity of a PRO measure (Patrick et al., 2011a,b). By doing so, content validity will lay the framework to subsequently aid in the interpretation of scores and in providing clarity for the communication of findings.

There are several types of qualitative research approaches, such as grounded theory, phenomenology, ethnography, case study, discourse analysis, and traditional content analysis; a comparison of these approaches can be found elsewhere (Lasch et al., 2010). The choice of approach will be dependent on the type of research question(s). However, for PRO development, the use of grounded theory is generally preferred (Kerr et al., 2010; Lasch et al., 2010).

Among the major facets of content validity is "saturation." Saturation refers to knowing when sufficient data have been collected to confidently state that the key concepts of importance for the particular patient group being studied have been captured. That is, if no new or relevant information is elicited, then there should be confidence that the main concepts of importance to patients and the items to measure them have been adequately obtained.

From the qualitative process, a draft of the conceptual framework emerges (see Figure 2.1 for an example), which is a visual depiction of the concepts, sub-concepts and items, and how they interrelate with one another. Often, the conceptual framework has been augmented by clinician input and a literature review in order to expand and refine the qualitative patient interviews. The hypothesized conceptual framework should be supported and confirmed with quantitative evidence.

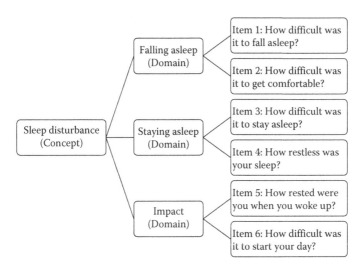

FIGURE 2.1
Example of a conceptual framework. (From Cappelleri, J.C. et al., *Patient-Reported Outcomes: Measurement, Implementation and Interpretation*, Boca Raton, Chapman & Hall/CRC Press, 2013.)

2.3 Construct Validity

Classical test theory (CCT) is a traditional quantitative approach to testing the reliability and validity of a scale based on its items, and is the basis for all of the psychometric methods described in this chapter (except for the person-item maps discussed in Section 2.6). In the context of PRO measures, CCT assumes that each observed score (X) on a PRO instrument is a combination of an underlying true score (T) on the concept of interest and unsystematic (i.e., random) error (E). CTT assumes that each person has a *true score* that would be obtained if there were no errors in measurement. A person's true score is defined as the expected score over an infinite number of independent administrations of the scale. Scale users never observe a person's true score, only an *observed score*. It is assumed that *observed score* (X) = *true score* (T) plus some *error* (E).

True scores quantify values on an attribute of interest, defined here as the underlying concept, construct, trait, or ability of interest (i.e., the "thing" that is intended to be measured). As values of the true score increase, responses to items representing the same concept should also increase (i.e., there should be a monotonically increasing relationship between true scores and item scores), assuming that item responses are coded so that higher responses reflect more of the concept.

CTT forms the foundation around construct validity. Constructs like hyperactivity, assertiveness, and fatigue (as well as anxiety, depression, and pain) refer to abstract ideas that humans construct in their minds in order to help them explain observed patterns or differences in their behavior, attitudes, or feelings. Because such constructs are not directly measurable with an objective device (such as a ruler, weighing scale, or stethoscope), PRO instruments are designed to measure these abstract concepts. A construct is an unobservable (latent) postulated attribute that helps one to characterize or theorize about the human experience or condition through observable attitudes, behaviors, and feelings (Cappelleri et al., 2013).

Construct validity can be defined as "the degree to which the scores of a measurement instrument are consistent with hypotheses (for instance, with regards to internal relationships, relationships with scores of other instruments, or differences between relevant groups)" (Mokkink et al., 2010). Construct validity involves constructing and evaluating postulated relationships involving a scale intended to measure a particular concept of interest. The PRO measure under consideration should indeed measure the postulated construct under consideration. If there is a mismatch between the targeted PRO scale and its intended construct, then the problem could be either that the scale is good but the theory is wrong, the theory is good but the scale is not, or that both the theory and the scale are useless or misplaced.

The assessment of construct validity can be quantified through descriptive statistics, plots, correlations, and regression analyses. Mainly, assessments of construct validity make use of correlations, changes over time, and differences between groups of patients. In what follows, the chief aspects of validity are highlighted (Cappelleri et al., 2013).

2.3.1 Convergent Validity and Divergent Validity

Convergent validity addresses how much the target scale relates to other variables or measures to which it is expected to be related, according to the theory postulated. For instance, patients with higher levels of pain might be expected to also have higher levels of depression, and this association should be sizeable. How sizeable? It depends on the nature of the variables or measures. Generally, though, a correlation between (say) 0.4 and 0.8 would seem reasonable in most circumstances as evidence for convergent validity (Cappelleri et al., 2013; de Vet et al., 2011; Fayers and Machin, 2016; Streiner et al., 2015). The correlation should not be too low or too high. A correlation that is too low would indicate that different things are being measured; a correlation that is too high would indicate that the same thing is being measured, and hence, that one of the variables or measures is redundant.

In contrast, *divergent* (or *discriminant*) *validity* addresses how much the target scale relates to variables or measures to which it is expected to have a weak or nonexistent relation (according to the theory postulated). For instance, little or no correlation might be expected between pain and intelligence scores.

As a validation method that combines both convergent validity and divergent validity, *item-level discriminant validity* can be conducted through tests involving corrected item-to-total correlations. Ideally, each item is expected to have a corrected item-to-total correlation of at least 0.4 with its domain total score (which is "corrected" for by excluding the item under consideration from its domain score). A domain, as defined here, is a subconcept represented by a score of an instrument that measures a larger concept consisting of multiple domains (FDA, 2009). Each item is expected to have a higher correlation with its own domain total score (after removing that item from the domain score) than with other domain total scores on the same questionnaire.

An example of convergent validity and divergent validity, as well as item-level discriminant validity, is found with the 14-item Self-Esteem And Relationship (SEAR) questionnaire, a 14-item psychometric instrument specific to erectile dysfunction (ED) (Althof et al., 2003; Cappelleri et al., 2004). Divergent validity on the eight-item Sexual Relationship Satisfaction domain of the SEAR questionnaire was hypothesized and confirmed by its relatively low correlations with all domains on the Psychological General Well-Being index and Short Form-36 (SF-36), both of which measure general health status. For the six-item Confidence domain of the SEAR questionnaire, divergent validity was hypothesized and confirmed by its relatively low correlations with physical factors of the SF-36 (Physical Functioning, Role-Physical, Bodily Pain, Physical Component Summary). Convergent validity was hypothesized and confirmed with relatively moderate correlations of the Confidence domain of the SEAR questionnaire and the SF-36 Mental Component Summary and the Role-Emotional and Mental Health domains, as well as with the Psychological General Well-Being Index (PGWBI) score and the PGWBI domains on Anxiety, Depressed Mood, Positive Well-Being, and Self-Control.

2.3.2 Known-Groups Validity

Known-groups validity is based on the principle that the measurement scale of interest should be sensitive to differences between specific groups of subjects known to be different in a relevant way based on accepted external criterion. As such, the scale is expected to show differences, in the predicted direction, between these known groups. The magnitude of the separation between known groups is more important than whether the separation is statistically significant, especially in studies with small or modest sample sizes in which statistical significance may not be achieved.

Consider that the known-groups validity of the SEAR questionnaire was based on a single self-assessment of ED severity (none, mild, moderate, severe) from 192 men (Cappelleri et al., 2004). Figure 2.2 contains the means and 95% confidence intervals for scores on the Sexual Relationship Satisfaction domain, the Confidence domain, and the 14-item Overall score of the SEAR questionnaire. For each, a score of 0 is least favorable, and a score of 100 is most favorable.

The mean scores across levels of ED severity differed significantly ($p = 0.0001$) and, as expected, increased (i.e., improved) approximately linearly

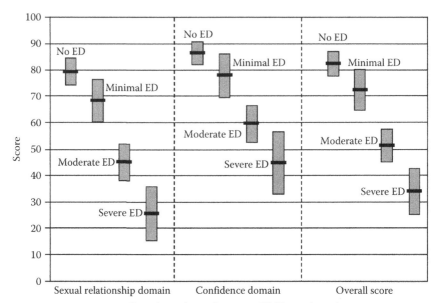

FIGURE 2.2
Mean scores and 95% confidence intervals on the SEAR questionnaire for four self-reported levels of ED. (From Cappelleri, J.C. et al., *Patient-Reported Outcomes: Measurement, Implementation and Interpretation*, Boca Raton, Chapman & Hall/CRC Press, 2013.)

and noticeably, with observed decreases in severity ($p = 0.0001$). Not only were each pair of mean scores statistically different ($p < 0.05$), the difference in mean scores on the SEAR questionnaire was present in the predicted direction and was tangible between each pair of adjacent categories on ED severity. The difference in mean scores between adjacent categories ranged from 8.4 (the difference in mean scores between no ED and mild ED on the Confidence domain) to 23.4 (the difference in mean scores between mild ED and moderate ED on the Sexual Relationship Satisfaction domain).

The *sensitivity* and *responsiveness* of a scale can be considered aspects (or at least, variations) of known-groups comparisons, and hence, of validity. *Sensitivity* is the ability of a scale to detect differences between patients or groups of patients when a difference is known to exist. While known-groups validity is concerned with confirming that anticipated differences are present between groups, sensitivity of a scale aims to detect statistical differences between treatments (or between other subdivisions of interest) that are of a magnitude considered to be meaningful or clinically relevant, and that aligns with expectations (Fayers and Machin, 2016; de Vet et al., 2011). A scale is useless for group assessment or evaluation unless it can reflect differences between patients with a poor prognosis and those with a good one, or between a beneficial treatment and a less beneficial treatment. If a treatment in a multiple-treatment clinical trial is known to be beneficial (relative to a control treatment) in terms of what the PRO is measuring, and the scale shows this in the way expected, the scale provides evidence of sensitivity in measuring the construct of interest.

Related to sensitivity, *responsiveness* refers to the ability of the scale to detect within-individual changes when a patient improves or deteriorates; for instance, responsiveness can be assessed between two groups using related samples of the same patients taken before and after an intervention. The responsiveness of a scale aims to detect statistical differences within patients (or a particular group of patients) of a magnitude considered to be meaningful or clinically relevant, and which also aligns with expectations. A highly sensitive scale will usually be a highly responsive one. Only measures with high test-retest reliability can detect real change and reduce the bias caused by measurement error. The use of a measurement scale is futile for patient monitoring unless it can reflect changes in a patient's condition.

It is known that statistical significance between groups and within groups is influenced in part by the selection of the patient sample and by the number of patients involved. Measures regarding the responsiveness of a scale should therefore also consider the magnitude of the intended effect and other items such as "effect size" (i.e., the difference between mean follow-up and mean baseline scores, divided by the standard deviation of scores at baseline) or standardized response mean (i.e., the difference between mean follow-up and mean baseline scores, divided by the standard deviation of the change scores) (Fayers and Machin, 2016), as an adjunct to (not a replacement for) the original (raw) scale metric, using the difference between means. Similarly, measures of the sensitivity of a scale, which should also consider the magnitude of the intended effect, include variants of effect size (e.g., the difference between mean score in one group and mean score in another group at follow-up, as divided by the pooled group standard deviation of scores at follow-up) as an adjunct to the scale in terms of its original metric, using the difference between means.

Such effect sizes provide a signal-to-noise ratio that is measured in standard deviation units and not in the original units of the PRO measure. However, because of this, effect size metrics are limited by a loss of direct clinical interpretation. In addition, confidence intervals should be used for quantifying the role of chance while stressing the magnitude of effect.

The responsiveness and sensitivity of different instruments can be evaluated by comparing their relative efficiency, which is based on the ratio of their test statistics (Fayers and Machin, 2016). Just like the hypothesis for testing construct validity, the hypothesis for testing responsiveness and sensitivity should include the expected direction (positive or negative) and the absolute or relative magnitude from the metrics used, be they correlations or differences between scores, or something else (de Vet et al., 2011).

2.3.3 Criterion Validity

Criterion validity involves assessing an instrument against the true value or against another standard that is indicative of the true value of measurement. It can be defined as "the degree to which the scores of a measurement instrument are an adequate reflection of a gold standard" (Mokkink et al., 2010). Criterion validity can be divided into concurrent validity and predictive validity.

Concurrent validity involves an assessment of scores from the targeted measure (e.g., the PRO measure of interest) with the scores from the gold standard at the same time. Despite there being a "suitable" criterion, there may be several valid reasons to develop and validate an alternative measure. These include that a current criterion measure may be too expensive, invasive, dangerous, or time-consuming; a current criterion may not have its outcomes known until it is too late; or the new measure may be measuring something similar but not exactly the same as what is taken or perceived as the gold standard. This last reason is generally why PRO measures are developed: because an existing instrument does not quite measure what is needed, despite it being taken or perceived previously as the gold standard.

An example of concurrent validity on a PRO instrument is found with the six-item erectile function domain on the 15-item International Index of Erectile Function (IIEF), where higher scores are more favorable (range: 1–30) (Rosen et al., 1997). Concurrent validity on the erectile function domain was examined, with the gold standard being a complete, documented clinical diagnosis of ED. For every one-point increase (benefit) in the erectile function score, the odds of having ED (relative to not having ED) decreased by about half.

Predictive validity involves an assessment of how well the targeted PRO measure predicts the gold standard in the future. An application of predictive validity is given together with three patient-reported quality-of-life questionnaires from the field of oncology, with progressive-free survival taken as the gold standard for this purpose (Cella et al., 2009). All three of these PRO measures at baseline are individually predictive of progression-free survival. The risk of tumor progression or death at any given time is approximately 7%, 11%, and 9% lower (depending on the PRO measure), respectively, for every unit of important benefit on the three patient-reported baseline measures.

2.4 EFA

Technically, factor analysis falls under the rubric of construct validity of instruments, but it has its own section here because of the technical exposition given to it. A factor is a latent or unobserved entity. A latent construct affects certain observed variables (or manifest variables) that can be measured directly. EFA (this section) and CFA (the next section, Section 2.5) are two major approaches to factor analysis, and several references are available for describing them in depth (Bollen, 1989; Brown, 2015; Cappelleri et al., 2013; Fayers and Machin, 2016; Kaplan, 2012; Kline, 2015; O'Rourke and Hatcher, 2013; Pett et al., 2003). Most statistical packages (such as SAS, R, SPSS, and Stata) are equipped to execute EFAs and CFAs.

2.4.1 Role of EFA

In EFA, there is an initial uncertainty as to the number of factors being measured, as well as which items are representing those factors. EFA is suitable for generating hypotheses about the structure of distinct concepts and regarding which items represent a particular concept. EFA can further refine an instrument by revealing which items may be dropped from the questionnaire, because they contribute little to the presumed underlying factors.

EFA is often confused with principal component analysis (PCA). Both EFA and PCA are variable reduction procedures, reducing a number of variables to a smaller number, and have been applied to determine item reduction and factor structure in the analysis of PROs. However, the two procedures are generally different and conceptually nonidentical. In EFA, the observed variables are linear combinations of the underlying factors, while in PCA, the principal components are linear combinations of the observed variables. Also, in EFA, factors are extracted to account only for the common item variance, while the unique variance remains unanalyzed.

In PCA, on the other hand, components are extracted to account for the total variance in the dataset, both common and unique, and not just the common variance. Because PCA makes no attempt to separate the common component from the unique component of each item's (variable's) variance, and EFA does, EFA is the appropriate method to use in identifying the factor structure of the data for PROs (and, more generally, in the behavioral and social sciences). The most widely used method in EFA to extract the factors is principal axis factoring, an approach that focuses on shared variance and not on sources of error that are unique to individual measurements. Another approach employed is the maximum likelihood method, which provides a significance test for solving the "number of factors" problem.

2.4.2 EFA Model

A factor is a latent or unobserved entity. A latent construct affects certain observed variables (or manifest variables) that can be measured directly. Figure 2.3 depicts a typical EFA model in which responses to questions 1

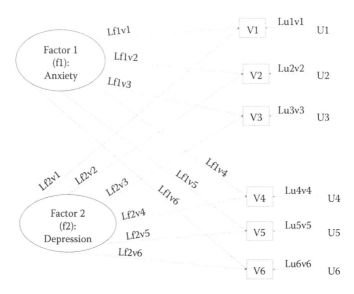

FIGURE 2.3
Illustrative EFA model. (From Cappelleri, J.C. et al., *Patient-Reported Outcomes: Measurement, Implementation and Interpretation*, Boca Raton, Chapman & Hall/CRC Press, 2013.)

through 6 are represented as six squares labeled V1 through V6. This path model suggests that there are two underlying factors, called Anxiety and Depression, but initially it is not known which items represent which domains. If variables V1, V2, and V3 are related and measure the same latent variable called Anxiety, then this construct (Anxiety) exerts a powerful influence on the way subjects respond to questions 1, 2, and 3 (notice the arrows going from the oval factor to the square variables). Similarly, the items V4, V5, and V6 are related to each other and influenced mainly by the second underlying factor, Depression. In Figure 2.3, only one factor is hypothesized to have a "substantial" loading for every variable; for example, V1 displays a substantial loading (Lf1v1) from Factor 1, but not from Factor 2 (Lf2v2 is not considered a meaningful or sizable loading).

The anxiety factor (Factor 1) in Figure 2.3 is known as a common factor, as is the depression factor (Factor 2). A common factor is a factor that influences more than one observed variable; it is a latent variable that is common to the set of manifest variables. Notice that in the EFA model in Figure 2.3, the two common factors (Anxiety and Depression) are not the only factors that influence the observed variables. For example, three factors actually influence variable V1: the common Factor 1, the common Factor 2, and a third factor labeled "U1." Here, U1 is a unique factor that only influences V1 and represents all of the independent and unmeasured factors that are unique to V1 (including the unique error component). The unique factor U1 affects only V1, U2 affects only V2, and so on. For variable V1, this relationship in the EFA model can be written as

$$V1 = Lf1v1 \times f1 + Lf2v1 \times f2 + Lu1v1 \times U1 \qquad (2.1)$$

In Figure 2.3, each of the arrows from a common factor to an observed variable is identified with a specific coefficient, such as Lf1v1, Lf1v2, Lf1v3, Lf1v4, Lf1v5, and Lf1v6 for the first factor f1. These coefficients represent factor loadings or path coefficients that embody the size of the effect that the underlying factor has on the variability of the scores of each observed variable. If the factor variables and observed variables are standardized to have unit variance (a variance of 1), those coefficients (loadings) are analogous to the standardized regression coefficients (regression weights) obtained in regression analysis.

2.4.3 Number of Factors

How does a researcher determine the number of factors that account for the relationships among the observed variables? That is, how many underlying dimensions (or subscales or domains) are there in the instrument? A widely used approach to determining how many domains or factors to retain is the scree test. The scree test is a rule-of-thumb criterion that involves the creation of a plot of the eigenvalues (i.e., the amount of incremental variance that is accounted for by a given factor) associated with each factor. The objective of a scree plot is to look for a break or dissociation between factors with relatively large eigenvalues and those with smaller eigenvalues; factors that appear before the break are taken to be meaningful and are retained. Researchers often refer to the break as the elbow in the curve of the scree plot.

Parallel analysis is another, more objective way to determine the number of factors to retain (Hayton et al., 2004; O'Connor, 2000). This method allows for the identification of factors that are beyond chance. Parallel analysis can be described as a series of three steps. The first step involves the generation of a random dataset with the same number of observations and variables as the real data being analyzed. Another part of this step is to randomly populate this dataset with values representing all possible response values of each item from the real dataset. In the second step, the newly generated random dataset is analyzed in order to extract the eigenvalues, with all eigenvalues saved for every simulation.

The first two steps should be repeated a sufficient number of times so as to create a stable distribution for every parallel eigenvalue. In the third step, the eigenvalues from the actual dataset are compared with the 95th percentile of the simulated eigenvalue. The first actual eigenvalue would be compared with the 95th percentile of the first random eigenvalue, the second actual eigenvalue would be compared with the 95th percentile of the second real eigenvalue, and so forth. The factors retained are only those whose eigenvalues in the actual dataset are greater than the 95th percentile of eigenvalues from the random data. An example of applying parallel analysis for a PRO is the Power of Food Scale (Cappelleri et al., 2009; see also the simulated example in Cappelleri et al., 2013).

In addition to the scree plot and parallel analysis, the following two other criteria can be used in the evaluation of the number of factors and their content for suitability: (1) items that load on a given factor should have a shared conceptual meaning; and (2) items should have high standardized factor pattern loadings (≥ 0.40 in absolute value) on one factor and low loadings on the other factors (O'Rourke and Hatcher, 2013).

2.4.4 Factor Rotation

A *rotation* is a transformation that is performed on the factor solution for the purpose of making the solution easier to interpret. In EFA, an orthogonal rotation is one in which the (common) factors are treated as uncorrelated, which is a very strong assumption and not a reasonable one to make for most PROs with more than one dimension; of the methods for orthogonal rotation, a varimax rotation is the most widely used. A more realistic approach is to use an oblique rotation, which allows for the factors to be correlated or associated, and, of the methods for oblique rotation, a promax rotation is the most widely used. This approach is realistic because the general expectation in measurement is that subscales of PROs tend to be at least somewhat correlated with each other, as they reflect underlying facets or aspects of a larger underlying construct. Thus, because factors tend to be correlated, an oblique rotation is often preferred over an orthogonal rotation. In particular, the use of a promax rotation makes relatively low variable loadings even lower by relaxing the assumption that factors should be uncorrelated with each other.

2.4.5 Sample Size

Factor analysis in general is a large-sample procedure, and a valid factor analysis typically involves hundreds of subjects. Regarding the minimum sample size needed for EFA, complete agreement among authorities of factor analysis is absent. For EFA, recommendations for the minimum number of subjects have ranged from 100 to 400 or more, depending on considerations such as the distribution of items and the correlations between them (Fayers and Machin, 2016). Some researchers have suggested that the minimum sample size should be at least five times the number of variables (or items) being analyzed. We prefer that, as a suggested rule of thumb, the sample size be at least 10 times the number of variables being analyzed. Therefore, for a 20-item questionnaire, at least 100 subjects and preferably 200 subjects would meet our suggested sample size for a study. As rough or crude approximations, this rule regarding the number of subjects per item constitutes a lower bound and thus may need to be modified depending on the amount of variability in the observed items, the number of items expected to load on each factor, and other considerations such as the magnitude of the correlations between the items.

2.4.6 Assumptions

The fundamental assumption underlying factor analysis is that one or more of the underlying factors can account for the patterns of covariation among a number of observed variables. Covariation exists when two variables vary together. Therefore, before conducting a factor analysis, it is important to analyze data for patterns of correlation. If no correlation exists, then a factor analysis is needless. If, however, at least moderate levels of correlation among variables are found, factor analysis can help uncover underlying patterns that explain these relationships.

EFA is generally intended for analyzing interval-scale data. An interval scale is one whose distance between any two adjacent points is the same. The Celsius temperature scale is an example of an interval measurement. The distance between 40°C and 41°C is exactly the same as the distance between 10°C and 11°C. Although factor analysis is designed for interval-scale data, many researchers also use the technique to analyze ordinal data. An ordinal scale is a ranking scale in which the differences between ranks are not necessarily equal. A Likert scale (e.g., strongly agree, agree, neither agree nor disagree, disagree, or strongly disagree) on which the responses are assigned a numerical value is an example of ordinal measurement.

Textbooks often state that, in principle, another assumption of EFA is that each observed variable be approximately normally distributed and, moreover, that each pair of observed variables involve approximately bivariate normal distribution. However, in practice, we have found that this assumption is not critical to the successful implementation of factor analysis. The Pearson correlation coefficient, the key driver of EFA, is robust against violations of the normal assumption when the sample size exceeds 25 (O'Rourke and Hatcher, 2013).

Tests for deciding on goodness-of-fit and the number of factors using maximum likelihood factor analysis, a less common variant of factor analysis, assume the multivariate normality that is required for significance tests.

Given that a factor-analytic model assumes that data are continuous (interval-level) and normally distributed, a natural question is whether factor analysis is well-suited for the analysis of PROs, where most variables are (strictly speaking) often neither continuous measures nor normally distributed. While more research is welcomed on the consequences of violating these two assumptions, empirical results and simulation studies suggest that factor analysis is relatively robust with regards to reasonable degrees of departure from interval-level data and normality (Fayers and Machin, 2016).

Factor analysis is a linear procedure of each observed item or variable regressed on a set of factors, so a linearity assumption is required as in the case of multiple linear regression.

2.4.7 Real-Life Application

Examples of EFA for PROs have been applied in diabetes, smoking cessation, urology, obesity, and in other diseases or therapeutic areas (Cappelleri et al., 2013; de Vet et al., 2011; Fayers and Machin, 2016; Streiner et al., 2015).

Consider, for instance, a version of the Minnesota Nicotine Withdrawal Scale (MNWS) for use in a smoking cessation trial. This version contains the following nine items: urge to smoke (item 1); depressed mood (item 2); irritability, frustration, or anger (item 3); anxiety (item 4); difficulty concentrating (item 5); restlessness (item 6); increased appetite (item 7); difficulty going to sleep (item 8); and difficulty staying asleep (item 9). Each item here is rated on a 0–4 ordinal response scale (0 = not at all, 1 = slight, 2 = moderate, 3 = quite a bit, and 4 = extreme).

The objective of the research was to identify the structure of this nine-item version of the MNWS and, in doing so, to refine and enhance its measurement properties of nicotine withdrawal symptoms. An EFA of the MNWS was conducted in a study ($n = 626$) across all available subjects at times at which varying levels of withdrawal symptoms were expected (week 0 [baseline], week 2, and week 4) (Cappelleri et al., 2005). This study was a Phase 2, multicenter, randomized, double-blind, parallel-group, placebo- and active-controlled study with a seven-week treatment phase. Subjects were randomized to one of three varenicline dose regimens (0.3 mg QD, 1.0 mg QD, or 1.0 mg BID); to the active control, sustained-release bupropion, 150 mg BID; or to placebo. The baseline or initial survey of subjects (week 0, prequit) was eight days before the target quit date (week 1 plus one day).

For the baseline data on the MNWS questionnaire, the scree plot depicted an abrupt break or discontinuity before eigenvalue 3, suggesting that only the first two factors were meaningful to be retained. An approximate straight line can be drawn from eigenvalue 3 to eigenvalue 9 throughout all points in between, but not from eigenvalue 2 or eigenvalue 1 to eigenvalue 9. The same pattern of eigenvalues was observed at week 2 and week 4.

Based on results of the rotated factor loadings (standardized regression coefficients) at different time points (baseline, week 2, and week 4) the following two multi-item domains emerged: Negative Affect, with four items (depressed mood; irritability, frustration, or anger; anxiety; difficulty concentrating), and Insomnia, with two items (difficulty going to sleep, difficulty staying asleep). In addition, three single items (manifest or observed variable: Urge to Smoke, Restlessness, and Increased Appetite), each measuring a distinct element of withdrawal, completed the remaining part of the MNWS structure.

2.5 CFA

2.5.1 EFA versus CFA

Like the purpose of EFA, the purpose of CFA is to examine latent factors that account for variation and covariation among a set of observed items or variables. Observed items (variables) are also known, in psychometric terminology, as indicators. Both types of factor analysis are based on the common factor model, and thus, they share many of the same concepts, terms, assumptions, and estimation methods. In CFA, as well as in EFA, factor loadings are coefficients that indicate the importance of a variable to each factor. These coefficients are important because they signify the nature of the variables that most strongly relate to a factor; the nature of the variables helps to capture the nature and meaning of a factor.

However, while EFA is generally an exploratory or hypothesis-generating procedure, CFA is usually a hypothesis-confirming technique that relies on a researcher's hypothesis and that requires prespecification of all aspects of the factor model, such as the number of factors and the patterns of indicator-factor relationships. While EFA explores the patterns in the correlations of items, CFA tests whether the correlations conform to an anticipated or expected scale structure given in a particular research hypothesis. Thus, CFA is suited in later phases of scale development or validation, after the underlying structure of the data has been tentatively established by prior empirical analyses using EFA, as well as by a theoretical understanding and knowledge of the subject matter.

As with structural equation modeling in general, CFA cannot prove causation per se. Rather, a major purpose of CFA is to determine whether the causal inferences of a researcher are consistent with the data; it shows that the assumptions made are not contradicted and may be valid. If the confirmatory factor model does not fit the data, then revisions are needed because then one or more of the model assumptions are incorrect and the measurement model thus requires refinement.

2.5.2 Measurement Model

Structural equation modeling is a comprehensive statistical approach to testing hypotheses about the relations among indicator (observed) and latent (unobserved) variables. Indicator variables are also considered manifest variables because, through them, only latent variables can manifest their current states. A structural equation model can be viewed as consisting of two components: (1) a measurement model that describes the relationships between the latent factors and their indicator variables; and (2) a structural model that describes the relationships between the latent factors themselves.

In the case of CFA, the structural part only generally specifies that latent variables are correlated, with no indication of causality. The measurement model not only covers a prespecified number of unobserved factors (latent variables) being assessed, but also covers a theoretical framework of which observable (manifest) variables are affected and not affected by which factors. A measurement model in a CFA can initially spring from an EFA or from a conceptual model based on subject matter knowledge. Within a structural equation modeling framework, variations exist regarding the structural or causal component around the common theme that involves tests for a specified relationship between concepts (constructs) of interest and, in doing so, allows for testing hypotheses on which factors affect other factors. One particular variation involves statistical mediation models (VanderWeele, 2015).

2.5.3 Standard Model versus Nonstandard Model

A standard CFA model refers to a system in which all variables constituting the structural portion of the model are latent factors with multiple indicators. Here, multiple manifest variables are used as indicators of each latent factor. It may not be possible, however, to include multiple measures for each construct in the structural model. It may be necessary to use a single indicator as part of a CFA because of the nature of the variable (e.g., vomiting) or the unavailability of multiple measures of a construct. This type of CFA model has some structural variables as single indicators and others as latent factors with multiple indicators, and is referred to as a nonstandard model.

2.5.4 Depicting the Model

Creating a diagram of the confirmatory factor model with its postulated latent variables and corresponding indicator variables is a natural way to begin a CFA. In Figure 2.4, we reconsider the same model described earlier in this chapter (see Figure 2.3). However, now Factor 1 only affects the indicator variables V1, V2 and V3, as represented by rectangles, and Factor 2 only affects the indicator variables V4, V5, and V6. The two factors are connected by a curved two-headed arrow, meaning that they are allowed to covary. This particular connection between factors is somewhat analogous to the oblique solution in EFA discussed earlier in this chapter (Section 2.4).

Figure 2.4 is an example of a basic standard model in which each indicator variable is assumed to be affected by only one factor. In Figure 2.4, notice also that no covariances exist between any of the indicators. The reason for this is that only constructs that are influenced solely by variables that lie outside of the model—that is, variables with no predictions made about them (like Factors 1 and 2 in Figure 2.4)—are allowed to have covariances. These variables are called exogenous variables. Variables predicted to be casually

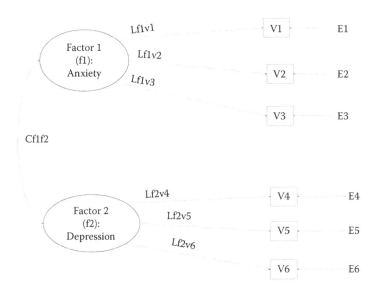

FIGURE 2.4
Illustrative CFA model. (From Cappelleri, J.C. et al., *Patient-Reported Outcomes: Measurement, Implementation and Interpretation*, Boca Raton, Chapman & Hall/CRC Press, 2013.)

affected by other variables, such as the indicators V1–V6 in Figure 2.4, are not allowed to have covariances. These variables are called endogenous variables.

2.5.5 Identifying Residual Terms for Endogenous Variables

A residual term for a variable represents all of the factors that influence variability in the variable, other than variables that precede and predict that variable. A residual term must be identified for each endogenous variable in the model and, because all of the indicator variables are endogenous variables (i.e., are affected by factor variables), a residual term must be created for each indicator. In Figure 2.4, the residual term begins with the letter "E" and ends with the same numerical suffix as its corresponding indicator. Thus the residual for V1 is E1, the residual for V2 is E2, and so forth. Each indicator is affected only by the underlying common factor on which it loads, along with a residual term. For example, indicator V1 is affected only by Factor 1 and E1, while V4 is affected only by Factor 2 and E4, and so on. This relationship, for variable V1, for instance, in the CFA model can be written as

$$V1 = Lf1v1 \times f1 + E1 \tag{2.2}$$

The comparison of Equation 2.2 with Equation 2.1 highlights the main methodological difference between EFA and CFA: specifically, in EFA, all factors potentially affect all indicator variables, but in CFA, only one factor is hypothesized to affect any indicator variable.

2.5.6 Identifying All Parameters to Be Estimated

Three types of parameters need to be estimated in Figure 2.4: variances of exogenous variables, covariances between the two factors, and factor loadings. Variances should be estimated for every exogenous (indicator) variable in the model. This means that the variance needs to be estimated for each E (residual) term in the model. Variances are not estimated for endogenous (variables) which, as a hypothetical construct (rather than a real-world observed variable), has no established metric or scale. Instead, the factors can be given unit variances by fixing their variances at 1.

Next, the covariance between the factors needs to be estimated. In Figure 2.4, the symbol "Cf1f2" represents this covariance estimate. Finally, the factor loadings need to be estimated. Factor loadings are basically path coefficients for the paths leading from a factor to an indicator variable. In Figure 2.4, the "L" symbol for "Loading" appears on the causal arrow from factor to indicator—for example, Lf1v1 represents the path coefficient from Factor 1 to V1, and Lf2v6 represents the path coefficient from Factor 2 to V6. Factor loadings are estimated for every causal path from factor to indicator. If the path coefficient (factor loadings) is relatively large and significantly different from zero, it means that the indicator is doing a good job with measuring the factor.

2.5.7 Assessing Fit between Model and Data

In CFA, a measurement model is postulated that predicts the existence of a specific number of latent factors and anticipates which indicator variables are affected by (the load on) each factor. The model is tested in a sample of subjects drawn from a population of interest. If the model provides a reasonable approximation to reality, the model should adequately account for the observed relations in the sample dataset—that is, the model should provide a "good fit" to the data.

A host of model-fit indexes exist. Among the most primary of them are the following: the Goodness-of-Fit Index, the Comparative Fit Index (CFI), the Normed Fit Index, the Non-Normed Fit Index, and the Root-Mean-Square Error of Approximation (RMSEA). For the first four of these indices, values above 0.90 generally indicate an acceptable fit (O'Rourke and Hatcher, 2013). For the RMSEA, values below 0.10 can be considered desirable, and 90% confidence intervals for the true RMSEA are often obtained (Steiger, 1999).

Tests of statistical significance for the unstandardized and standardized factor loadings should also be conducted. Unstandardized factor loading can be interpreted in the same way as unstandardized regression coefficients, where the metric of the indicator variable is passed onto the latent factor, so that a one-unit increase in the factor is associated with a given change in the indicator. Also relevant are the magnitudes of the standardized factor loadings, where the metrics of both the indicators and factors are standardized to have a mean of 0 and a variance of 1, and their interpretation is the same as standardized regression coefficients. Thus they reflect an increase in one standardized score in the factor that is associated with a given change in one standardized score in the indicator. Being measured in standard deviation units, standardized factor loadings with values of 0.40 or larger (in absolute values) can be considered noteworthy.

Some researchers also report a model chi-square statistic (the null hypothesis of the perfect model fit). The model chi-square statistic with a low enough p-value (say, less than .05) suggests rejection of the null hypothesis of perfect fit. This hypothesis, however, is likely to be implausible because it is unrealistic to expect a model to have perfect fit. Moreover, the chi-square statistic is known to increase its sensitivity and get larger (and hence, be more probable to lead to the rejection of the null hypothesis), with larger correlations and larger sample sizes, even for a very good fitting model in which the differences between the observed and predicted covariances are slight.

Regarding the sample size required for CFA, rules of thumb have been offered including a minimum number of subjects per each parameter to be estimated (e.g., at least 10 subjects per parameter). The same caution given about such rules of thumb for EFA also applies to CFA. It is safe to say that, in general, hundreds of subjects would be needed. Statistical power and sample sizes for CFA are explained in more detail elsewhere (Brown, 2015).

Among the features of CFA is the ability to compare and test nested models, akin to the comparison and evaluation of nested models in regression analysis. For example, one model may allow the latent factors to be correlated, while another model may assume that they are orthogonal and not correlated. The resulting likelihood ratio test, which is the difference between the chi-square test results between the fit of the model where the latent factors are allowed to correlate (the parent model) and the model that constrains them to zero (the nested model), can be performed to compare the two models regarding whether the null hypothesis of no correlation between factors can be rejected in favor of an alternative hypothesis of nonzero correlation between at least one pair of factors.

2.5.8 Real-Life Application

Examples of CFA for PROs have been applied in smoking cessation, urology, obesity, and other disease or therapeutic areas (Cappelleri et al., 2013; de Vet et al., 2011; Fayers and Machin, 2016; Streiner et al., 2015). Reconsider, for instance, the smoking cessation application using the MNWS discussed in Section 2.4.7. Consequent to EFA from a Phase 2 study, a CFA using data from two Phase 3 studies (Gonzales et al., 2006; Jorenby et al., 2006) confirmed the hypothesis that this MNWS scale should be structured as two multi-item domains and three manifest variables, as suggested by the EFA. Values of the CFI exceeded 0.90, indicating that this model has an acceptable fit to the data, and other goodness-of-fit statistics were consistent with these findings. In addition to the fit indices, the factor loadings of the individual items on their respective factors were acceptably high, and were consistent across measurement points. In all instances, factor loadings are larger than 0.4, indicating that the items loaded solidly on their respective factors.

2.6 Person-Item Maps

CTT has been the mainstay of conventional psychometric analysis. Over the last few decades, however, psychometricians have advanced another measurement model: item response theory (IRT). IRT is a statistical theory consisting of mathematical models that express the probability of a particular response to a scale item as a function of the (latent or unobserved) attribute of the person and of certain parameters or characteristics of the item (Cappelleri et al., 2013; Embretson and Reise, 2000; Hambleton et al., 1991; Reeve, 2003). Rather than replacing CTT methodology, IRT methodology can be more constructively viewed as a potentially important complement to CTT for scale development, evaluation, and refinement, in certain circumstances.

In the context of health measurement, IRT assumes that patients at a particular level of an underlying attribute (e.g., with a particular level of physical functioning) will have a certain probability of responding positively to each question. This probability will depend, in part, on the difficulty of the particular item. For example, many patients with cancer might respond "Yes" to easy questions such as "Do you dress yourself?," but only patients with a high level of physical functioning are likely to reply "Yes" to the more difficult question "Do you engage in vigorous activities such as running, lifting heavy objects, or participating in strenuous sports?"

One popular type of IRT model is the Rasch model, which requires only the single difficulty parameter b_i to describe an item i (e.g., running would be considered to be more difficult than walking) (Andrich, 2011; Bond and Fox, 2015; Boone et al., 2014; Fischer and Molenaar, 1995). The probability of a positive response by a particular patient j to item i [$P_{ij}(\theta_j)$] is a function of the difference between the amount of the patient's latent attribute θ_j and the item's difficulty b_i. The difficulty parameter (b_i) in relation to the attribute parameter (θ_j) indicates the extent of a positive response to a particular item.

Three assumptions underlie the successful application of the Rasch model (and IRT models in general): unidimensionality, local independence, and correct model specification. The assumption of unidimensionality requires that a scale consists of items that tap into only one dimension. Local independence means that, for a subsample of individuals who have the same level on the attribute, there should be no correlation among the items.

Correct model specification, which is not unique to Rasch (or IRT) models, is necessary at both the item level and person level. For details on these assumptions and how to evaluate them, the reader is referred elsewhere (Bond and Fox, 2015; Embretson and Reise, 2000; Fischer and Molenaar, 1995; Hambleton et al., 1991).

At least five factors influence sample size estimation for Rasch models: (1) the type of response options (a larger number of categories requires a larger sample size); (2) the study purpose (a more definitive study requires a larger sample size); (3) the sample distribution of respondents (less uniformity or dispersion of responses requires a larger sample size); (4) the number of items (more items require larger sample sizes); and (5) the relationship between the set of items and the attribute of interest (a lesser relationship requires a larger sample size) (Cappelleri et al., 2013). Furthermore, more generally, the choice of IRT model affects the required sample size, with Rasch models generally requiring a smaller sample size than other IRT models. In general, Rasch models tend to require at least 200 patients for stable parameter estimation.

It is common in Rasch analysis to center the metric around item difficulty (instead of around person attribute), so that the mean of the item difficulties is fixed to equal 0. If the PRO measure is easy for the sample of persons, the mean across person attributes will be greater than zero ($\theta > 0$); if the PRO measure is hard for the sample, the mean of θ will be less than zero. Rasch users produce person-item (or Wright) maps to show such relationships between item difficulty and person attributes (Bond and Fox, 2015; Stelmack et al., 2004).

Figure 2.5 portrays such a person-item map of a 10-item scale on physical functioning. Because of the scale content, person attribute is referred to more specifically as person ability. With a recall period of the past four weeks, each item is pegged to a different physical activity, but raises the same question: "In the past four weeks, how difficult was it to perform the following activity?" Each item also has the same set of five response options: 1 = extremely difficult, 2 = very difficult, 3 = moderately difficulty, 4 = slightly difficult, and 5 = not difficult. Assume that all of the activities discussed in the questionnaire were attempted by each respondent. Also assume that item difficulty emanated from the rating scale model, a polytomous Rasch model, where each item has its own difficulty parameter separated from the common set of categorical threshold values across items (Bond and Fox. 2015).

At least four points are noteworthy. First, the questionnaire contains more easy items than hard ones, as 7 of the 10 items have location (logit) scores on item difficulties below 0. A consequence of this is that a score jumps from 1 (item on "Walking > one mile") to slightly below 3 ("Vigorous activities," such as running, lifting heavy objects, or participating in strenuous sports), which suggests that adding some moderately difficult items would be beneficial in order to measure and distinguish among individuals within this range.

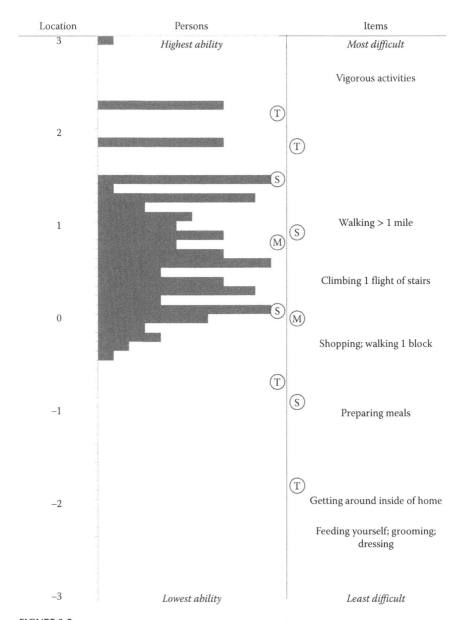

FIGURE 2.5
Illustration of person-item map on physical functioning. (From Cappelleri, J.C. et al., *Patient-Reported Outcomes: Measurement, Implementation and Interpretation*, Boca Raton, Chapman & Hall/CRC Press, 2013.)

Note: M = mean of person distribution or item distribution, S = single standard deviation from the person mean or the item mean, T = two standard deviations from the person mean or the item mean.

Second, some items have the same difficulty scores: the dual items on shopping and walking one block; and the triplet on feeding yourself, grooming, and dressing. Not much scale information would be sacrificed if one of the dual items and two of the triplet items were removed.

Third, if the items have been written based on a construct map (a structured and ordered definition of the underlying ability or attribute intended to be measured by the scale and conceived of in advance), then the item map (e.g., the right-hand side of Figure 2.5) that follows the construct map can be used as evidence congruent with content validity. A construct map is informed by a strong theory of which characteristics require higher levels of the attribute for endorsement.

Finally, patients tend to cluster at the higher end of the scale (note that the mean location score is about 1 for the ability of persons, and it exceeds the fixed mean location of 0 for the difficulty of items), indicating that most patients would be likely to endorse (or respond favorably to) several of these items. Figure 2.5 shows which items fall toward the lower end (most persons would endorse or respond favorably to these items) and which items rose to the upper end (relatively few persons would endorse or respond favorably to these items). Thus, this group of patients had either a high degree of physical functioning or, consistent with the previous evaluation of the items, there are not enough challenging or more difficult items, such as those regarding moderate activities (such as moving a table, pushing a vacuum cleaner, bowling, or playing golf), present in the questionnaire. The ideal situation is to have a mean location score of around 0 ($\theta = 0$) across a representative sample of individuals.

2.7 Reliability

2.7.1 Repeatability Reliability

As part of scale validation, two standard types of reliability for PROs are *internal reliability* for multi-item scales and *repeatability reliability* (Cappelleri et al., 2013). Both concepts are mathematically related. Internal reliability, also known as internal consistency, is based on item-to-item correlations and the number of items in multi-item scales. Repeatability reliability, which is applicable to single-item scales as well as to multi-item scales, is based upon the analysis of variances between repeated measurements on the same set of subjects, where the measurements are repeated over time (*test-retest reliability*).

Regarding repeatability or test-retest reliability, if a patient is in a stable condition under practically identical circumstances, an instrument should yield reproducible results when it is repeated on that patient. Under this condition, when patients complete the same PRO questionnaire on multiple occasions (time points), the level of agreement between the PRO scores at multiple occasions is a measure of the reliability of the PRO measure.

For PRO data taken as continuous, the *intraclass correlation coefficient* (ICC) is a reliability parameter that measures test-retest reliability and the strength of agreement between repeated measurements on the same set of patients by assessing the proportion of total variance (with total variance being the sum of between-patient variance plus within-subject variance) in the observed measurements that is due to true differences between patients as reflected in the between-patients variability in the observed scores. The theoretical range on an ICC is from 0 to 1, with higher values indicating more reliability. ICCs are available for either a single measurement or an average of multiple measurements.

As an example, consider the Erection Hardness Score (EHS), a single-item PRO measure on a five-point scale (0 = penis does not enlarge to 4 = penis is completely hard and rigid) that measures erection hardness for men being treated for their ED (Mulhall et al., 2007). Test-retest analyses were based on EHS data from event logs (at each event of sexual intercourse) collected, before treatment intervention, during the two-week screening phase, and at baseline. The ICC was 0.51 for a single EHS response. As expected, reliability increased with the number of EHS responses, and the cutoff for acceptable reliability (≥ 0.7) was attained when three EHS responses were averaged, which gave an estimated test-retest reliability of 0.76.

In practice, the Pearson correlation and the intraclass correlation tend to be quite close. Nonetheless, the intraclass correlation is preferred as it, unlike the Pearson correlation, will yield a value of 1.0 only if all of the observations on each patient are identical; the Pearson correlation can yield a value of 1.0 even when all of the observations on each patient are not identical.

For categorical data, the *kappa statistic* is a measure of agreement between two nominal variables having two or more categories (Cohen, 1960). The kappa coefficient explicitly adjusts for agreement that occurs by chance alone, and thus can be defined as chance-corrected agreement. For example, the data in Table 2.2 on the diagnosis of ED come from two methods: a gold standard (clinical) diagnosis and a PRO-based diagnosis (erectile function domain of the IIEF). The overall agreement observed is simply $(1000 + 102) / 1151 = 0.96$ (or 96%). After correction for the expectation that a certain number of agreements would arise by chance alone, the kappa coefficient (or the proportion of chance-corrected agreement) between the erectile function domain of the IIEF and the clinical diagnosis was 0.78, which is substantial given that the maximum value of kappa is 1.

The simple kappa coefficient only considers total or full agreement and does not provide partial credit or agreement for responses that are in proximity or nearby, for instance, in that they differ by only one or two categories for variables with more than two levels or categories. An extension of (simple) kappa, *weighted kappa*, does consider partial agreement for ordinal variables (Cohen, 1968). For instance, the weighted kappa was 0.80 between the

TABLE 2.2

Cross-Tabulation of Erectile Function Scores (Pro Measure) and Clinical Diagnosis (Gold Standard)

Erectile Function Domain	Gold Standard		Total
	Clinical Diagnosis of ED	Clinical Diagnosis of No ED	
ED (≤ 25)	1000 (true positive)	14 (false positive)	1014
No ED (26–30)	35 (false negative)	102 (true negative)	137
Total	1035	116	1151

ED = Erectile Dysfunction.
Source: Cappelleri, J.C. et al., *Patient-Reported Outcomes: Measurement, Implementation and Interpretation*, Boca Raton, Chapman & Hall/CRC Press, 2013.

true grade (severity) of erectile function and the predictive grade of erectile function (Cappelleri et al., 1999). Men were classified correctly with respect to their "true" level of ED more often than any other level. Men who were misclassified tended to be assigned into a degree category adjacent to the correct or true category, rather than to a more remote category.

Simple kappa and weighted kappa can be interpreted as follows: ≤0, poor; 0–0.2, slight agreement; 0.21–0.40, fair agreement; 0.41–0.60, moderate; 0.61–0.80, substantial; and 0.81–1.00, almost perfect (Landis and Koch, 1977). Kappa has also been interpreted with less than 0 being no agreement and 0–0.2 being poor agreement (Elliott and Woodward, 2009).

2.7.2 Internal Consistency Reliability

Internal consistency reliability applies to the consistency of responses to items on the same multi-item scale, where the items are intended to tap into the same construct (or complementary interrelated aspects of it). This form of reliability uses item correlations to assess the homogeneity (similarity) of items on a multi-item scale at a particular time, and refers to the extent to which the items are interrelated.

Cronbach's alpha coefficient is the most widely used method to assess internal consistency (Cronbach, 1951). Values of Cronbach's alpha are between 0 and 1. These values can be calculated in two ways: correlation-based formula and the covariance-based formula. The correlation-based formula incorporates the average (mean) correlation between all pairs of items and the number of items. The higher the mean inter-item correlation and the number of items, the higher Cronbach's alpha is.

Unlike the correlation-based formula, which assumes that each item be standardized to have the same variance, and weighs each item equally, the covariance-based formula provides a raw (unstandardized) score that allows items to have different variances. Here, items with larger variances will have greater weight in the computation of Cronbach's alpha in the raw score formula. Both methods for computing Cronbach's alpha tend to be quite similar in practice. Which of the two formulas is preferred depends on the specific context and whether equal weighting is desired.

A well-accepted guideline for the value of Cronbach's alpha is between 0.70 and 0.90 (Cappelleri et al., 2013). However, this general recommendation is also meant to be placed into perspective, taking into account the length of the scale and the set of inter-item correlations.

As an example, in an observational study of 192 men (98 who reported a clinical diagnosis of ED in the past year and 94 who reported no clinical diagnosis of ED in the past year), Cronbach's alpha was determined for the SEAR questionnaire (Cappelleri et al., 2004). Cronbach's alpha for the Sexual Relationship Satisfaction domain, the Confidence domain, and Overall score were 0.91, 0.86, and 0.93, respectively. Cronbach's alpha for the Self-Esteem subscale and Overall Relationship Satisfaction subscale of the Confidence domain were 0.82 and 0.76, respectively.

2.8 Conclusions

Developing the content of a new patient-reported measure takes much care, time, and consideration to minimize error in measuring the concept that you are interested in measuring (the concept of interest). Not only do we want to be sure that we are measuring what we say we are measuring, we also want to reduce noise or ambiguity.

This chapter covers in part the different facets of validity, or measuring what is intended to be measured. Content validity is the extent to which an instrument measures the important aspects of concepts that developers or patients purport it to assess. Construct validity is the degree to which the scores of a measurement instrument are consistent with hypotheses. Different aspects of construct validity are considered: convergent validity and divergent validity, whereby a PRO scale is assessed in relation to other variables to which it is expected to be related or not related; and known-groups validity, whereby a PRO scale is assessed with respect to differences between specific groups of subjects known to be different in a relevant way. Criterion validity, the third type of validity, involves assessing a PRO scale against the true value or against some other standard indicative of the true value of measurement. Two types of criterion validity—concurrent validity and predictive validity—are addressed.

As a model-based form of validity, factor analysis is discussed. The purpose of factor analysis is to examine latent factors that account for variation and covariation among a set of observed items or variables. EFA is generally an exploratory or hypothesis-generating procedure. In contrast, CFA is usually a hypothesis-confirming technique that relies on a researcher's hypothesis and that requires prespecification of all aspects of the factor model. While EFA explores the patterns in the correlations of items, CFA tests whether the correlations conform to an anticipated or expected scale structure given in a particular research hypothesis.

Additional insights about validity can be obtained through a person-item map, which has its roots in Rasch methodology. In this chapter, a person-item map is illustrated to show the relationship between the difficulty of items and the level or location that persons have on the attribute of interest. These maps can illuminate the extent of item coverage or comprehensiveness on the attribute of interest, the amount of item redundancy, and the range and location of the attribute in the sample.

This chapter also covers reliability—measuring accurately, with precision, what is intended to be measured. As one form of reliability, repeatability (test-retest) reliability is described as a way to examine and quantify the degree to which measurement is stable and free from measurement error. As another form of reliability, internal consistency reliability is considered as a way to quantify the consistency of responses to items on the same multi-item scale.

Validity is limited by reliability. If responses are inconsistent (unreliable), then it necessarily implies invalidity as well. Note that the converse may not be true: consistent responses do not necessarily imply valid responses. Good evidence of both validity and reliability is needed in order to ensure that a PRO measure is measuring what it is intended to measure, and is doing so with precision. Such evidence is especially relevant and warranted in a regulatory environment where the intention is for a sponsor of a medicinal product to seek a label claim granted by a regulatory authority.

While this chapter does not cover the regulatory landscape, Chapter 8, which is on the reporting, dissemination, and interpretation of results, does. Generally, findings measured by PROs may be used to support claims in approved medical product labeling when the claims are derived from adequate and well-controlled investigations in which PROs measure specific concepts accurately and as intended. Such PROs can be developed and assessed in accordance with regulatory guidance documents from the Food and Drug Administration (FDA, 2009, 2010) and the European Medicines Agency (EMA, 2005, 2009).

Acknowledgments

This chapter is based in part on the material in Chapters 2 through 6 of our monograph, Cappelleri, J.C. et al., *Patient-Reported Outcomes: Measurement, Implementation and Interpretation*. Boca Raton, FL: Chapman & Hall/CRC Press, 2013. In addition, we are grateful and appreciative to Linda Deal and an anonymous reviewer for helpful and constructive comments, which improved the quality of the exposition.

References

Ahmed, S., Berzon, R.A., Revicki, D.A. et al. 2012. The use of patient-reported outcomes (PRO) within comparative effectiveness research: Implications for clinical practice and health care policy. *Med Care* 50:1060–1070.

Alemayehu, D., Sanchez. R.J. and J.C. Cappelleri. 2011. Considerations on the use of patient-reported outcomes in comparative effectiveness research. *J Manag Care Pharm* 17:S27–S33.

Althof, S.E., Cappelleri, J.C., Shpilsky, A. et al. 2003. Treatment responsiveness of the Self-Esteem And Relationship (SEAR) questionnaire in erectile dysfunction. *Urology* 61:888–893.

Andrich, D. 2011. Rating scales and Rasch measurement. *Expert Rev Pharmacoecon Outcomes Res* 11:571–585.

Basch, E., Abernethy, A.P., Mullins, C.D. et al. 2012. Recommendations for incorporating patient-reported outcomes into clinical comparative effectiveness research in adult oncology. *J Clin Oncol* 30:4249–4255.

Bollen, K.A. 1989. *Structural Equations with Latent Variables*. New York, NY: John Wiley & Sons.

Bond, T.G. and C.M. Fox. 2015. *Applying the Rasch Model: Fundamental Measurement in the Human Sciences*. 3rd ed. Mahwah, NJ: Lawrence Erlbaum Associates.

Boone, W.J., Staver, J.R. and M.S. Yale. 2014. *Rasch Analysis in the Human Sciences*. New York, NY: Springer.

Brown, T.A. 2015. *Confirmatory Factor Analysis for Applied Research*. 2nd ed. New York, NY: The Guilford Press.

Cappelleri, J.C., Althof, S.E., Siegel, R.L. et al. 2004. Development and validation of the Self-Esteem And Relationship (SEAR) questionnaire in erectile dysfunction. *Int J Impot Res* 16:30–38.

Cappelleri, J.C., Bushmakin. A.G., Baker, C.L. et al. 2005. Revealing the multidimensional framework of the Minnesota nicotine withdrawal scale. *Curr Med Res Opin* 21:749–760.

Cappelleri, J.C, Bushmakin, A.G., Gerber, R.A. et al. 2009. Evaluating the power of food scale in obese subjects and a general sample of individuals: Development and measurement properties. *Int J Obes* 33:913–922.

Cappelleri, J.C., Rosen, R.C., Smith, M.D. et al. 1999. Diagnostic evaluation of the erectile function domain of the International Index of Erectile Function. *Urology* 54:346–351.

Cappelleri, J.C., Zou, K.H., Bushmakin, A.G. et al. 2013. *Patient-Reported Outcomes: Measurement, Implementation and Interpretation*. Boca Raton, FL: Chapman & Hall/CRC Press.

Cella, D., Cappelleri, J.C., Bushmakin, A. et al. 2009. Quality of life predicts progression-free survival in patients with metastatic renal cell carcinoma treated with sunitinib vs. interferon-alfa. *J Oncol Pract* 5:66–70.

Cohen, J. 1960. A coefficient of agreement for nominal scales. *Educ Psychol Meas* 20:37–46.

Cohen, J. 1968. Weighted kappa: Nominal scale agreement with provision for scaled disagreement or partial credit. *Psychol Bull* 70:213–220.

Cronbach, L.J. 1951. Coefficient alpha and the internal structure of tests. *Psychometrika* 16:297–334.

de Vet, H.C.W., Terwee, C.B., Mokkink, L.B. et al. 2011. *Measurement in Medicine: A Practical Guide*. New York, NY: Cambridge University Press.

Doward, L.C., Gnanasakthy, A. and M.G. Baker. 2010. Patient reported outcomes: Looking beyond the claim. *Health Qual Life Outcomes* 8:89. Open access.

Elliott, A.C. and W.A. Woodward. 2009. *SAS Essentials: A Guide to Mastering SAS for Research*. San Francisco, CA: Jossey-Bass.

Embretson, S.E. and S.P. Reise. 2000. *Item Response Theory for Psychologists*. Mahwah, NJ: Lawrence Erlbaum Associates.

European Medicines Agency (EMA), Committee for Medicinal Products for Human Use. 2005. *Reflection paper on the regulatory guidance for us of health-related quality of life (HRQOL) measures in the evaluation of medicinal products*. European Medicines Agency. http://www.ema.europa.eu/ema/index.jsp?curl=pages/regulation/general/general_content_000366.jsp (accessed May 31, 2017).

European Medicines Agency (EMA), Committee for Medicinal Products for Human Use. 2009. *Qualification of novel methodologies for drug development: Guidance to applicants*. European Medicines Agency. http://www.ema.europa.eu/ema/index.jsp?curl=pages/regulation/document_listing/document_listing_000319.jsp (accessed May 31, 2017).

Fayers, P.M. and D. Machin. 2016. *Quality of Life: The Assessment, Analysis and Reporting of Patient-Reported Outcomes*. 3rd ed. Chichester, UK: John Wiley & Sons.

Fischer, G.H. and I.W. Molenaar, eds. 1995. *Rasch Models: Foundations, Recent Developments, and Applications*. New York, NY: Springer.

Food and Drug Administration (FDA). 2009. Guidance for industry on patient-reported outcome measures: Use in medical product development to support labeling claims. *Fed Regist* 74(235):65132–65133.

Food and Drug Administration (FDA). 2010. Draft guidance for industry on qualification process for drug development tools. *Fed Regist* 75(205):65495–65496.

Gonzales, D., Rennard, S.I., Nides, M. et al. 2006. Varenicline, an alpha4beta2 nicotinic acetylcholine receptor partial agonist, vs sustained-release bupropion and placebo for smoking cessation: A randomized controlled trial. *JAMA* 296:47–55.

Hambleton, R.K., Swaninathan, H.J. and H.J. Rogers. 1991. *Fundamentals of Item Response Theory*. Newbury Park, CA: Sage Publications.

Hayton, J.C., Allen, D.G. and V. Scarpello. 2004. Factor retention decisions in exploratory factor analysis: A tutorial on parallel analysis. *Organ Res Methods* 7:191–205.

Jorenby, D.E., Hays, J.T., Rigotti, N.A. et al. 2006. Efficacy of varenicline, an alpha-4beta2 nicotinic acetylcholine receptor partial agonist, vs placebo or sustained-release bupropion for smoking cessation: A randomized controlled trial. *JAMA* 296:56–63.

Kaplan, D. 2012. *Structural Equation Modeling: Foundations and Extensions*. 2nd ed. Thousand Oaks, CA: Sage Publications.

Kerr, C., Nixon, A. and D. Wild. 2010. Assessing and demonstrating data saturation in qualitative inquiry supporting patient-reported outcomes research. *Expert Rev Pharmacoecon Outcomes Res* 10:269–281.

Kline, R. 2015. *Principles and Practice of Structural Equation Modeling*. 4th ed. New York, NY: The Guilford Press.

Landis, J.R. and G.G. Koch. 1977. The measurement of observer agreement for categorical data. *Biometrics* 33:159–174.

Lasch, K.E., Marquis, P., Vigneux, M. et al. 2010. PRO development: Rigorous qualitative research as the crucial foundation. *Qual Life Res* 19:1087–1096.

Mokkink, L.B., Terwee, C.B., Patrick, D.L. et al. 2010. International consensus on taxonomy, terminology, and definitions of measurement properties for health-related patient-reported outcomes: Results of the COSMIN study. *J Clin Epidemiol* 63:737–745.

Mulhall, J.O., Goldstein, I., Bushmakin, A. et al. 2007. Validation of the erectile hardness score. *J Sex Med* 4:1626–1634.

O'Connor, B.P. 2000. SPSS and SAS programs for determining the number of components using parallel analysis and velicer's MAP test. *Behav Res Meth Instrum Comput* 32:396–402.

O'Rourke, N. and L. Hatcher. 2013. *A Step-by-Step Approach to Using the SAS® System for Factor Analysis and Structural Equation Modeling*. 2nd ed. Cary, NC: SAS Institute.

Patrick, D.L., Burke, L.B., Gwaltney, C.H. et al. 2011a. Content validity–Establishing and reporting the evidence in newly developed patient reported outcomes (PRO) instruments for medical product evaluation: ISPOR PRO good research practices task force report: Part 1–Eliciting concepts for a new PRO instrument. *Value Health* 14:967–977.

Patrick, D.L., Burke, L.B., Gwaltney, C.H. et al. 2011b. Content validity–Establishing and reporting the evidence in newly developed patient reported outcomes (PRO) instruments for medical product evaluation: ISPOR PRO good research practices task force report: Part 2–Assessing respondent understanding. *Value Health* 14:978–988.

Pett, M.A., Lackey, N.R. and J.J. Sullivan. 2003. *Making Sense of Factor Analysis: The Use of Factor Analysis for Instrument Development in Health Care Research*. Thousand Oaks, CA: Sage Publications.

Reeve, B.B. 2003. Item response theory modeling in health outcomes measurement. *Expert Rev Pharmacoecon Outcomes Res* 3:131–145.

Rosen, R.C., Riley, A., Wagner, G. et al. 1997. The international index of erectile function (IIEF): A multidimensional scale for assessment of erectile dysfunction. *Urology* 49:822–830.

Steiger, J.H. 1999. *EzPATH: Causal Modeling*. Evanston, IL: SYSTAT.

Stelmack, J., Szlyk, J.P., Stelmack, T. et al. 2004. Use of a Rasch person-item map in exploratory data analysis: A clinical perspective. *J Rehabil Res Dev* 41:233–242.

Streiner, D.L., Norman, G.R. and J. Cairney. 2015. *Health Measurement Scales: A Practical Guide to their Development and Use*. 5th ed. New York, NY: Oxford University Press.

VanderWeele, T.J. 2015. *Explanation in Causal Inference: Methods for Mediation and Interaction*. New York, NY: Oxford University Press.

3

Observational Data Analysis

Demissie Alemayehu, Marc Berger, Vitalii Doban, and Jack Mardekian

CONTENTS

3.1 Introduction 47
3.2 Confounding in Causal Inference 48
3.3 Design and Analytical Considerations 50
    3.3.1 Design Options 50
    3.3.2 Regression Adjustment Analysis Methods 52
    3.3.3 PS Analysis 52
    3.3.4 Instrumental Variables 55
    3.3.5 Other Considerations 56
        3.3.5.1 Sensitivity Analysis 56
        3.3.5.2 Structural Equation Models 56
        3.3.5.3 Time-Dependent Covariates 57
        3.3.5.4 Recent Developments 57
3.4 Operational Considerations 57
    3.4.1 Data Warehousing and Processing 58
    3.4.2 Data Standards 59
    3.4.3 Computing and Data Visualization 60
    3.4.4 Security and Privacy 60
3.5 Best Practices for the Analysis and Reporting of
    Observational Studies 61
3.6 Real-World Data in Application 62
3.7 Concluding Remarks 63
References 64

3.1 Introduction

It is well known that randomized controlled trials (RCTs) hold a prominent place in the hierarchy of evidence generation. This is mainly because of the fact that randomization ensures the minimization of bias emanating from a lack of comparability of treatment groups with regards to potential confounding factors. On the other hand, reliance on RCTs may not always be tenable, since such trials may not be feasibly conducted, or the information provided by them may not be adequate to make important health care decisions. Under such circumstances, it may be essential to use information from observational studies. There also seems to be a growing recognition of the importance of such data by regulators (Sutter, 2016).

On the flip side, the primary weakness of observational studies is the issue of confounding resulting from imbalances among treatment groups, in terms of known and unknown confounding factors that may impact the outcome of interest (Deeks et al., 2003). Further, in contrast to data from RCTs, there are issues of standards and quality that may constrain the accessibility and utility of information from observational studies (Alemayehu and Mardekian, 2011).

In view of the considerable potential of real-world data in making critical health care decisions, there have been concerted efforts to maximize the value of such data, both in terms of addressing methodological challenges, as well as in establishing best practices for executing studies (Berger et al., 2014; Cox et al., 2009; Johnson et al., 2009; von Elm et al., 2008).

In this chapter, we highlight the major statistical and infrastructural issues and, in doing so, provide a summary of best practices developed for the effective use of real-world data. The rest of this chapter is organized as follows: in Section 3.2, we discuss confounding in the context of nonrandomized studies. Section 3.3 outlines commonly used design options and analytical strategies. In Section 3.4, operational considerations are addressed. In Section 3.5, we review guidelines to enhance the analysis and reporting of observational studies. Section 3.6 includes selected examples from the relevant literature, while Section 3.7 provides concluding remarks.

3.2 Confounding in Causal Inference

A confounder is technically defined as a variable that is associated with the response as well as the treatment. In general, it is not possible to identify all possible confounders within a study. Thus, in the absence of randomized assignment, the comparability of the study treatment groups with respect to potential confounders cannot be ascertained. Further, even in situations in which important confounders could be listed, there may not have been adequate data collected to permit cogent analysis. In certain special cases such as in *confounding by indication*, which is common in drug safety studies in which the indication is also a risk factor for the outcome, confounding cannot be completely removed by design or modeling when no control exists for the underlying condition (Bosco et al., 2010; Psaty and Siscovick, 2010).

The concept of counterfactual causality plays an important role in the development of a framework for statistical inference pertaining to causal effects (Heckman, 2005). To fix ideas, we wish to evaluate the effect of a given treatment T relative to a control C with respect to an outcome Y. Let Y_{it} denote the response of an individual i to receiving treatment t ($t = T$ or C). It is noted that for a given i, Y can be observed only under either $t = T$ or C, but not both. The *potential outcome* for the subject i corresponding to the treatment that the subject has not actually received is referred to as *counterfactual*. In this framework, the magnitude of causal effect for the subject i may be computed as

$$\delta_i = Y_{it} - Y_{ic}$$

Given a representative sample of n subjects, the population *average causal effect*, where effect is quantified in terms of means, may then be estimated based on

$$\bar{\delta} = \frac{1}{n}\sum_{i=1}^{n}\delta_i$$

A key supposition in causal inference is the so-called *stable-unit-treatment* assumption, which implies that the outcome of individual i under $t = T$ or C is independent of the responses of other individuals or their treatment assignments. Another common assumption is that of *exchangeability*, which signifies that the distribution of the unobserved Y_t for $t = T$, when $t = C$ is the same as that of the observed Y_T (i.e., under the actual treatment $t = T$) and vice versa (i.e., when the roles of C and T are reversed). The concept of *exchangeability* is relevant in the choice of optimal study designs and estimators that are unbiased. For example, in the above setting in which the effect is expressed as a mean, if there are no other sources of bias, then the *average causal effect* can be directly estimated as the difference in means:

$$\bar{\delta} = \bar{Y}_T - \bar{Y}_C$$

In a subsequent section, we will introduce the important concept of the *propensity score* (PS), which denotes the probability that an individual is assigned to a given treatment. In the special case of simple randomized trials, these scores are expected to be equal for all individuals. On the other hand, in an observational study, where the group assignment is not under the control of the investigator, the PS may depend on variables that may or may not be observable.

It may be noted that, despite its widespread use, the counterfactual approach has remained a center of controversy. Notably, Dawid (2000) criticized the counterfactual concept on the grounds that causal inference is dependent on unobservable assumptions. Rothman (1976) proposed the so-called sufficient-component-cause model, in which units of the model are not individuals but are instead mechanisms of causation (with each mechanism being a combination of factors that are jointly sufficient to induce a binary outcome event). Nonetheless, the prevailing view is that the counterfactual framework is useful in addressing fundamental and practical issues arising in the analysis of data from observational studies and, as such, is widely embraced by researchers in the field.

3.3 Design and Analytical Considerations

The statistical literature is replete with alternative approaches to minimize the effects of confounders in observational studies (Rosenbaum and Rubin, 1983; Waning and Montagne, 2001). Understandably, the choice of an analytical procedure should generally be dependent on the study design, and on the availability of data on confounders of interest. For example, when the design involves matching, the selected procedure should take into account the accompanying correlation induced by the matching mechanism. Recent studies have shown that results of observational studies tend to vary widely as a function of trial design and analytical procedures employed (Madigan et al., 2013a). In the following section, we summarize pertinent aspects of some of the commonly used design options and methods of analysis.

3.3.1 Design Options

In a *cohort study*, groups of study participants that use the drug of interest and others that use a suitably chosen comparator are prospectively identified based on predefined criteria. The response is then compared in the two groups, using models that adjust for relevant confounders. This design permits the performance of an analysis of outcome measures on both categorical and continuous scales. Typical effect measures that may be analyzed include mean differences, risk ratios, odds ratios, and other quantities commonly used in epidemiological studies. Cohort study designs may be executed as either prospective or retrospective trials. Prospective cohort studies tend to be resource-intensive and generally require a lengthy amount of time for data collection. However, such studies are often appealing since they can be used to address several diseases concurrently. On the other hand, retrospective cohort studies can be relatively less costly to complete as compared with prospective cohort studies, but may be limited by the availability of data for analysis (Kleinbaum et al., 2013).

In situations in which the outcome of interest is binary, one often conducts case-control studies, which typically involve starting with considering cases and noncases (controls) of a disease or other health outcome, and subsequently proceeding backwards to determine prior exposure history. In matched *case-control* designs, subjects having a given outcome (cases) are matched with those without the outcome (controls) according to a prespecified matching criterion, and the rates of exposure in the two groups are then compared. Analysis is often performed using procedures that take into account the dependence induced by the matching mechanism. These designs are appealing primarily because of cost- and time-related reasons, as they are cheaper and less time-consuming to perform than prospective cohort studies. However, their shortcomings include susceptibility to selection bias, a limitation in not allowing the direct estimation of relative risks, and infeasibility for consideration of rare exposures (Kleinbaum et al., 2013). Further, their results may lack generalizability, since subjects in the study sample are selected according to the outcome values. A recent study reported that such studies tend to give biased estimates (Madigan et al., 2013c).

Although cohort and case-control studies are common, other design options are also available that may prove useful under certain conditions. When the case is a rare event and the risk factor of interest is expensive to measure, one may employ a *case-cohort* study, in which the control is sampled at baseline, or a *nested case-control* design, in which the control is sampled at the time when Y is observed. In both designs, the cases used can serve as their own controls.

In a case-cohort design, which combines features of both case-control and cohort designs, controls are sampled from the original cohort, while the included cases are new or incident events of a disease. These designs tend to be more prone to measurement error than a typical cohort study, and more costly than a case-control study.

In a nested case-control study, which is a variation of the case-cohort design, controls are matched to the cases at the time of diagnosis (referred to as density sampling). Typically, one or more controls are selected for each case from subjects in the original cohort who are still at risk at the time a case is identified. Controls for a given case may later become cases after the time they are selected as controls. While such designs may be more susceptible to measurement error than cohort studies, one potential advantage in their favor is that controls are at risk for a comparable amount of time as the matched cases (Kleinbaum et al., 2013). However the analytical strategy should take into account aspects of the design, especially when subjects serve as their own control. In a purely *self-controlled* study, outcomes are compared before and after exposure in the same subject. An advantage of this design is that covariates that don't vary within a person during the study period are automatically controlled. Typical analytical techniques under these conditions include conditional logistic and Poisson regression models, depending on the outcome measures (e.g., all occurrences versus only the first occurrence of an event).

Variations of the above designs include the *self-controlled cohort design*, in which post-exposure incidence rates are compared with pre-exposure incidence rates among patients exposed to the target drug of interest. In this design, an external comparator group is not used, unlike in cohort designs, and the analysis is not restricted to within-person comparisons (see, e.g., Madigan et al., 2013c).

Lastly, *cross-sectional* studies are often used to obtain estimates of the prevalences of diseases. This design, which is relatively convenient and inexpensive to use, involves taking a random sample at a given point in time, and permits handling multiple outcomes and exposures in real time. The results, however, can only be used for the purpose of hypothesis generation. The design cannot establish whether the exposure preceded the disease and, in addition, may under-represent diseases with short durations (Kleinbaum et al., 2013).

3.3.2 Regression Adjustment Analysis Methods

When data is available on relevant covariates, traditional models, including analysis of covariance, generalized linear models, or Cox proportional hazards models, could be used, depending on the outcome variable, to adjust for confounding effects. Under valid assumptions, these procedures are generally known to be optimal, and to provide results that are relatively easy to understand and interpret. On the other hand, they can be sensitive to departure from model assumptions, and may perform poorly under nonideal conditions, including in the presence of severe multicollinearity or influential points. Further, when the number of covariates is large relative to the size of the study subject cohort, they yield inefficient estimators and may even breakdown. To address the issues of high dimensionality and multicollinearity, researchers have used tools that involve regularization. Examples of such techniques include ridge regression and the least absolute shrinkage and selection operator (LASSO), which are discussed in Chapter 4 (see also Hastie et al., 2009).

3.3.3 PS Analysis

Originally introduced in Rosenbaum and Rubin (1983), the PS technique, alluded to earlier, involves computing subjects' likelihood of receiving treatment as a function of prespecified risk factors. To fix ideas, suppose we have two treatment groups, denoted by Z, having a value of 1 if the subject is exposed, and 0 otherwise. The PS for an individual is then the conditional probability of being treated, given the covariates:

$$p_i = Pr(Z = 1 \mid \text{covariates for subject } i)$$

Implicit in the approach is the idea that the framework creates a "quasi-randomized" experiment, such that subjects with comparable p_i tend to have similar distributions in the covariates. In this regard, the method is dependent on the covariates included, and cannot handle the impacts of latent confounders, except to the extent that unmeasured covariates are associated with those included in the calculations of the p_is.

Once PSs are estimated, they can be used in 1:1 or *M:1* matching, in which subjects are grouped together based on their *PS* values (D'Agostino, 1998). For example, in a 1:1 assignment, the so-called *greedy matching* assigns to a randomly selected exposed subject an unexposed subject with the closest PS value, with the process repeated until all of the exposed subjects are matched. Alternatively, an optimal criterion may be specified in which the aim is to minimize the total within-pair difference of the scores. As noted in Gu and Rosenbaum (1993), *optimal matching* may be comparable to greedy matching in its ability to produce balanced matched samples. In *nearest-neighbor* matching, the scores of matched subjects must be below some prespecified threshold (the *caliper distance*). In this approach, if there are no untreated subjects within the caliper distance, then the treated subject under consideration is excluded from the resultant matched sample. Although there is no standard rule for defining maximal distance, matching using a caliper width of 0.2 times the standard deviation of the logit of the PS has been recommended (Austin, 2009a). When analyses are performed based on PS matching, it is debatable whether the analysis needs to account for the matched pair nature of the data. One position advocates that the model used should take into account the matched nature of the sample (Austin, 2009c), whereas another position advocates reasons for not accounting for the matched pairs and for analyzing the two groups independently (Stuart, 2010).

An example of a study that used PS matching was recently published by Andersen et al. (2016). Patients in the test group were matched with those in the control using nearest neighbor matching with a maximum caliper of 0.01 on the PS scale. The PS was calculated using multivariable logistic regression with generalized estimating equations to account for clustering within hospitals. The two groups of patients in the propensity matched cohort were well balanced with respect to all included variables.

The PS is often incorporated in the analysis stage using other alternative approaches, including stratification, covariate adjustment, or inverse probability weighting. In the first case, subjects are stratified into disjoint subsets based on prespecified thresholds. A common convention is to divide subjects into five equal-size groups with reference to the quintiles of the estimated PSs (Cochran, 1968). The analysis may then be performed by pooling across stratum-specific estimates, or by applying standard techniques, such as traditional analysis of variance, logistic regression, or Cox proportional hazards models, in which the PS strata are included as a stratification term in the model. A major limitation of the stratification approach is the inevitable loss of power and precision due to exclusion, if each treatment group is not adequately represented in each stratum.

In certain applications, the estimated PS is included as one of the covariates in the model for the purpose of covariate adjustment. However, this approach has been shown to yield biased estimates (Austin et al., 2007a). Consequently, alternative remedial measures have been proposed (see, e.g., Imbens, 2004).

The *inverse probability of treatment weighting* (IPTW) method involves assigning each subject a weight equal to the inverse of the probability of receiving the treatment that the subject actually received. Using the previous notation (Z_i) for treatment, the weight for the individual i is given by

$$w_i = \frac{Z_i}{p_i} + \frac{(1 - Z_i)}{(1 - p_i)}$$

Given outcome Y, the average treatment effect δ is estimated by

$$\hat{\delta} = \frac{1}{n} \sum_{i=1}^{n} \frac{Z_i Y_i}{p_i} - \frac{1}{n} \sum_{i=1}^{n} \frac{(1 - Z_i)Y_i}{1 - p_i}$$

Lunceford and Davidian (2004) offer an estimate of the standard error of $\hat{\delta}$. A limitation of the IPTW approach is that the use of weights may be unstable for subjects with small values of the PS (Robins et al., 2000).

Detailed discussions of the relative performances of the above procedures may be found in Austin (2007) and Austin et al. (2007b). It is noted, however, that, despite their popular use, PS approaches may not always be suitable. In one study, for example, it was shown that the use of PS methods, while advantageous in certain settings, did not demonstrate a substantial benefit as compared with conventional multivariable methods (Sturmer et al., 2006). Further, PS matching is only as good as the variables included in the analysis. While there is no general approach to selecting variables for inclusion, clinical knowledge should play a critical role in screening the pool of candidate variables.

PS analysis generally requires the use of large samples to attain distributional balance of observed covariates and, for the method to be effective, there must be substantial overlap between the treatment groups on the PSs. It is therefore essential to conduct extensive sensitivity analyses to ensure that balance is achieved with respect to the individual covariates across the treatment groups in subjects with comparable PS. One recommended criterion of balance is that the absolute standardized difference of the baseline characteristics in matched patients be less than 10% (Austin, 2009b).

When there are more than two groups in a study, it may be difficult to extend the above approach using multinomial models. When there is a reference control group, it is customary to perform pairwise matching with the control. For example, in a recently published retrospective study involving three test drugs with a control, PS matching was performed separately in three pairs of matched cohorts, in which each pair contained the control (Yao et al., 2016). A review of the baseline characteristics table in the article indicates adequate matching by pairs that might not have been as readily achievable by the use of overall matching based on a multinomial approach.

Lastly, traditional methods used to compute PS are known to be susceptible to the problem of high dimensionality and collinearity. Recently, the so-called *high-dimensional PS* (hd-PS) has been proposed as a viable option, in which a large number of diagnoses, medications, and procedure codes are ranked according to their likelihood of confounding. However, the approach is prone to subjectivity, and seems to rely heavily on the idea that a large number of "proxy" variables can reduce bias from unmeasured confounding (Schneeweiss et al., 2009). Alternatively, modern analytic tools, some of which do not rely on model specifications, and that generally work well in high-dimensional cases, can be used in estimating the PSs. Examples of such techniques proposed in the context of PS include neural nets (Setoguchi et al., 2008) and ensemble methods, such as random forests (Lee et al., 2010).

3.3.4 Instrumental Variables

Instrumental variables (IVs), which were originally developed for use in economic models, are now commonly applied to minimize the effects of unmeasured confounders in observational studies (Newhouse and McClellan, 1998). The choice of an IV is generally nontrivial, and there is no straightforward way to identify an instrument when one actually exists. In any case, an IV should satisfy certain key assumptions. First, an IV should be strongly associated with the treatment, essentially by sharing a common cause. However, the IV should not share a common cause with the outcome variable. It should be related to the outcome only through its association with the treatment (i.e., it is conditional on treatment and other covariates, it has no effect on the outcome). Ideally, it should also be unrelated to the observed and unobserved patient characteristics that affect the outcome (i.e., it behaves like a factor that is randomly assigned to the groups under comparison).

In the search for a proper IV, one needs to identify a quasi-random treatment assignment. One common example relates to *interruptions in medical practice*, which may be a consequence of changes following a regulatory requirement or an important innovation. Another involves *treatment preference*, independent of patient factors. Instances of the latter may include distance to specialists (McClellan et al., 1994), geographic areas (Stukel et al., 2007), physician prescribing preference (Brookhart et al., 2006), and hospital formulary (Schneeweiss et al., 2007). In a recent study on the association between lymphadenectomy performance and patient survival, Wright et al. (2016) used geographic variation in the execution of lymphadenectomy as an IV.

A common approach in IV analysis is the so-called two-stage least-squares method, which entails first predicting the unmeasured propensity for assignment to treatment (T) using the IV and other covariates, followed by estimating the outcome based on the predicted treatment and measured covariates. More specifically, in matrix notation, denoting the IV and vector of covariates by V and X, respectively, the system of equations is given by (where the superscript t represents transpose on the column vector of coefficients)

$$T = \alpha_0 + \alpha_1^t X + \varphi V + \varepsilon_1$$

$$Y = \beta_0 + \beta_1^t X + \gamma T + \varepsilon_2$$

One drawback of IV approaches is that the underlying assumptions are not readily verifiable. However, every effort should be made to authenticate the validity of the assumptions empirically whenever the available data permits. For example, the association between the IV and treatment, or its association with measured confounders, could be evaluated using standard statistical techniques. Similarly, whether there is no direct relationship between the IV and the outcome other than through actual treatment may be assessed by formulating a suitable model.

3.3.5 Other Considerations

3.3.5.1 Sensitivity Analysis

Sensitivity analysis is a useful tool to evaluate the robustness of study results to plausible scenarios of bias, including robustness of a given design relative to assignment bias (Rosenbaum, 2004). Conventional sensitivity analysis, however, depends on the values of the parameters considered. Alternatively, Monte Carlo sensitivity analysis, which assigns distributions to the unknown bias parameters, may be a more attractive option (Greenland, 2005).

3.3.5.2 Structural Equation Models

Structural equation models (SEMs) are commonly used in causal modeling. A key feature of these models is the inclusion of parameters pertaining to the relations among the latent variables, as well as those among latent and observed variables. In application, the use of SEMs requires caution, since the approach often involves complex assumptions, which are often difficult to justify. Most of the assumptions are essential to ensure parameter identifiability. In some instances, the coefficients in the models may not be interpretable, or the meaning of the latent variables may not be obvious. A discussion of these models may be found in McCallum and Austin (2000). For a causal interpretation of SEMs and some of the misconceptions concerning the approach, see also Bollen and Pearl (2013). For an example of SEM use in causal modeling to investigate the factors that affect hospitalization costs in community-acquired pneumonia patients, see Uematsu et al. (2015).

3.3.5.3 Time-Dependent Covariates

When a covariate of interest is not constant but varies over time for a given individual, statistical inference for causal effects tends to be rather complex. In cases where the outcome variable is survival time, an analysis may be performed using suitable survival models. For complex cases involving measured and unmeasured confounders, it may be essential to use nonstandard approaches, including those proposed in Robins et al. (1992).

3.3.5.4 Recent Developments

As mentioned above, recent developments in the machine learning literature have offered additional tools for use in the analysis of real-world data. Most of the approaches are algorithmic in nature and require little or no model specification. However, an appealing feature of these techniques is their ability to fairly effectively handle two of the common problems that arise in the analysis of real-world data, namely high dimensionality and multicollinearity (Hastie et al., 2009). Although there is still a tendency among data analysts to use traditional statistical models for the analysis of real-world data, there seems to be a growing body of literature reporting results based on modern analytic tools (Marino et al., 2012). A more detailed discussion of the approaches is given in Chapter 4 in this monograph.

3.4 Operational Considerations

The effective use of data from observational studies entails addressing major infrastructural challenges, which are no less important than the methodological and conceptual issues discussed above. Health care data may come from different sources (e.g., electronic medical records [EMRs] or claims databases), use different formats (e.g., text, numeric, images, etc.), or require large storage spaces. As a result, prior to their use for research purposes, it is essential to have a reliable and efficient computing infrastructure, including hardware, software and support staff, available. In the following, we highlight some of the operational issues, with particular reference to data warehouses, computing environments, data standards, and the protection of privacy and confidentiality. For a more detailed discussion, the reader should consider Alemayehu and Mardekian (2011).

3.4.1 Data Warehousing and Processing

In order to facilitate analyses, a first step in the provisioning of health care data is to ensure that the disparate data can be aggregated into a single, central system. Commonly used systems for this purpose include enterprise data warehouses (EDWs) that host all the source data "as is" and specialized datamarts that host data for specific therapeutic areas or certain other specific patient cohorts, to enable more focused data analyses and better performance. While EDWs are usually implemented as on-premise (a.k.a. on-prem) solutions (i.e., the EDW is hosted behind a company's firewall), in recent years, cloud computing has become a viable option for managing such data via facilitating the storage, access, and processing of data online. In contrast to the traditional on-prem EDW model, cloud computing is highly parallelized, which means that large jobs can be run in parallel across many nodes. One big advantage that allowed cloud computing to be widely adopted by large and small companies alike is the ability to scale on-demand, that is, to easily provision small or large data environments depending on the current needs in place and without large upfront investments, as is necessary in the case of EDWs. On-demand provisioning and large-scale cloud computing providers can also lead to significant cost savings. However, one perceived concern with cloud computing is data security. This has initially slowed down the adoption of cloud computing solutions, especially in industries that deal with highly sensitive data, such as health care. However, as cloud computing offerings become more mainstream and continue to demonstrate improved reliability, cost savings, and performance as compared with on-prem EDW solutions, they are likely to achieve increasingly wider adoption.

Historically, many of the data preparation steps were performed by software developers, rather than by data analysts. With the ever-increasing need for speed in generating data insights, and with data being more accessible to frontline analysts, a new generation of powerful and user-friendly data preparation tools have evolved over the past few years. These visually rich, intuitive, drag-and-drop tools now allow less technically savvy users easy access to data and the ability to perform many data preparation tasks on their own, including data look-ups, data validation, missing values detection/replacement, and combining data from multiple heterogeneous data sources. Further, the tools facilitate building collaborative data preparation workflows that can implement end-to-end repeatable data pipelines, thereby allowing data analysts to be more self-sufficient.

The new generation of data management tools is especially relevant in the context of EMR and claims data, which require extensive data cleaning and processing. Different EMR systems use different nomenclatures for various medical terms and codes (e.g., diagnosis codes with or without the decimal point), different coding conventions or measurement units for observations, and lab results (e.g., gender coded as M/F or as 0/1, weight and height stored in the metric or standard system). Also, because a sizable portion of EMR data is manually entered by busy health care providers, such as physicians and nurses, data entry errors are common and that leads to many data issues, such as data completeness and physiologically impossible values (e.g., negative LDL/HDL values).

Obvious data entry errors are further complicated by decisions regarding how to address them. For example, a body mass index value of 321 could very well result from incorrect data entry of 32.1, but analysts may be reluctant to "correct" the value for analysis and instead may prefer to reassign the value as missing. On the other hand, a 12-month health care cost of $1 million for a patient diagnosed with a common condition like the occurrence of migraines may be an error, but the replacement value may not be intuitive. Further, missing values are ubiquitous in such data. An important statistical assumption that is made in handling them is that they are missing at random. However, in observational studies, the missingness mechanism may be informative. For example, lipid level readings may not be taken if the physician has no underlying reason to order their collection.

3.4.2 Data Standards

Most health record data used in observational studies are collected in a non-uniform manner. In some cases, the data may be unstructured, while in others, they may be structured. It is also not uncommon to see inconsistent or variable definitions of data elements. Thus, the lack of standards poses a considerable challenge to the analyst, limiting the ability to efficiently process the data and apply routine statistical software.

Even with the most advanced and user-friendly tools, it takes a lot of effort to implement data cleaning, standardization, and analytical procedures across multiple data sources. One approach to addressing this issue is to build a Common Data Model (CDM), which entails a data structure that allows for the loading and storing of different data sources into a standard data structure. Some researchers and institutions build their own CDMs, while others tend to use one of the publicly available CDMs, such as the Sentinel CDM (SCDM) or the OMOP (Observational Medical Outcomes Partnership) CDM. Most of the common CDMs not only standardize the data structures and harmonize the vocabularies across different sources that allow for the storing of them in a common data structure, but also enable collaboration among researchers and the development of tools and analytical approaches that can be applied across different data sources.

There are also concerted efforts to harmonize certain types of health care data. Examples include the Clinical Data Interchange Standards Consortium (CDISC)'s Healthcare Link project, which was conceived with the goal of harmonizing data from health care and clinical research (CDISC, 2017), and the *International Classification of Diseases, Ninth Revision, Clinical Modification* (ICD-9-CM), used to code and classify diagnoses from inpatient and outpatient records. More recently, there is the enhanced *International Classification of Diseases, Tenth Revision, Clinical Modification* (ICD-10-CM). In other cases, prescription drugs and insulin products are coded using the National Drug Code (NDC) scheme that is maintained by the United States Food and Drug Administration (FDA, 2015). More recently, there has been a growing movement to harmonize data collection across states (Porter et al., 2015). However, despite the growing awareness of the need for data standardization, much work is still needed to harmonize the collection, coding, and storage of such data.

3.4.3 Computing and Data Visualization

In tandem with effective data warehousing and standardization, it is crucial to have high performance analytical tools, especially for data visualization and exploratory purposes, on hand. Historically, observational datasets were used by a limited number of sophisticated researchers who used specialized statistical software to answer carefully constructed and predefined analyses and research questions. Today, these datasets are used to answer a much broader spectrum of questions posed by a larger community of researchers and business users from the pharmaceutical industry, academia, government, and health care providers and payers. As a result, powerful analytic tools that can be used by less sophisticated researchers to query and analyze the data are desired.

Preferred tools offer user-friendly and easy-to-use interfaces and allow for the building of powerful visualizations that are not only aesthetically pleasing and easier to interpret than tabular output, but also enable the incorporation of a lot of flexibility using embedded run-time filters. These filters should enable users to change analysis parameters "on the fly," without having to rerun the analysis from scratch. For example, users should be able to change the patient cohort they are analyzing by easily filtering according to patient age or gender, presence of a diagnosis or a medication, or enrollment periods, using options built into the software.

With the growing interest in predictive analytics and data mining, choices of customized software to perform data visualization and execute simple and more complex analyses are now in abundance. However, caution should be exercised in the application of these tools to generate evidence with expected impacts on health care, since appropriate use may require a thorough understanding of the operating characteristics of the underlying statistical procedures.

3.4.4 Security and Privacy

A major concern with the use of health care data is the need to ensure the integrity of the data and the privacy of patients. In this regard, every effort should be made to ensure that the infrastructure used for the collection, storage, and access of such data has adequate safeguards to guarantee data security and the protection of patient privacy. In terms of the latter, at a minimum, this entails adherence to applicable laws, including the Health Insurance Portability and Accountability Act of 1996 (HIPAA, Title II) in the U.S. and its enhancement, the 2009 Health Information Technology for Economic and Clinical Health (HITECH) Act, which significantly expanded the scope of HIPAA's privacy rules.

The issue is particularly germane to health records handled through the cloud computing paradigm. As reported in one study, to guarantee the protection of the information and to adequately comply with privacy policies, concrete steps have to be taken by both the cloud computing service providers and their health care customers, including with respect to access privileges, network security mechanisms, data encryption, digital signatures, and access monitoring (Rodrigues et al., 2013).

The interplay among computing efficiency, data privacy, and classical statistical inference has also garnered increased attention in the statistical and computing literature. See Duchi et al. (2013) and Wainwright (2014) for a discussion of approaches in which privacy is incorporated as a constraint on the data analytic model in a classical statistical decision-theoretic framework.

3.5 Best Practices for the Analysis and Reporting of Observational Studies

With the growing attention on the value of data from observational studies in evidence-based medicine, numerous guidelines have been developed to promote good research practices in this area. Key elements of the guidelines include the need to specify the research hypothesis, study design, and analytical approaches in a study protocol; the emphasis on the importance of sensitivity analyses to ensure the robustness of the study findings; and the reporting of the results with fair balance, highlighting the limitations of the study and the generalizability of the results.

Examples include the recommendations of the International Society for Pharmacoeconomics and Outcomes Research (ISPOR) (Berger et al., 2009; Cox et al., 2009; Johnson et al., 2009); the Agency for Healthcare Research and Quality (AHRQ) (Velentgas et al., 2013); the International Society of Pharmacoepidemiology (2008); the STrengthening the Reporting of OBservational studies in Epidemiology (STROBE) statement (von Elm et al., 2008); Good Research for Comparative Effectiveness (GRACE) principles (Dreyer et al., 2010); and other resources for evaluating nonrandomized studies of comparative effectiveness (Deeks et al., 2003; Schneeweiss and Avorn, 2005; Tooth et al., 2005).

STROBE consists of a checklist of 22 items that relate to the title, abstract, results, and discussion sections of articles. Under the Methods section, there are items that address study design, setting, participants, variables, data sources/measurement, bias, study size, quantitative variables, and statistical methods. The latter addresses issues of confounding, subgroups and interactions, missing data, and sensitivity analyses.

The related guidance, "A Questionnaire to Assess the Relevance and Credibility of Observational Studies to Inform Health Care Decision Making: An ISPOR-AMCP-NPC Good Practice Task Force Report" (Berger et al., 2014), consists of 33 items divided into two domains: relevance and credibility. The relevance domain consists of four questions addressing the population, interventions, outcomes, and context.

The credibility domain consists of subsections that address study design, data, analyses, reporting, interpretation, and conflicts of interest. Under the Design subsection, it asks whether study hypotheses or goals were prespecified a priori, whether comparison groups were concurrent, whether there existed a formal study protocol including an analysis plan prior to executing the study, what the sample size and power calculations are, whether the design minimized or accounted for confounding, the appropriateness of the follow-up period, the appropriateness of sources/criteria/methods for selecting participants, and the similarity of comparison groups. Under the Data subsection are questions related to the sufficiency of the data sources, the validity of how exposure was defined and measured, the validity of how primary outcomes were defined and measured, and whether the follow-up across comparison groups was comparable. Under the Analysis subsection are questions that address potential measured and unmeasured confounders, subgroups and interactions, and sensitivity analyses. Under the Report subsection are questions related to how the final sample was defined, descriptive statistics of study participants, description of key components of statistical analyses, the reporting of confounder-adjusted estimates of treatment effects, description of statistical uncertainty, missing data, and the reporting of absolute and relative treatment effects. Under the Interpretations subsection, there are questions about whether the results are consistent with known prior information, whether the results are clinically meaningful, whether the conclusions are supported by the data and analyses, and whether the effect of unmeasured confounding was discussed.

3.6 Real-World Data in Application

The reliability of data from observational studies in evidence-based medicine has been the focus of extensive debate in the scientific literature (Concato and Horwitz, 2004). Over the years, several studies have been conducted to evaluate whether observational studies provide results that are consistent with those of RCTs. Examples of cases in which positive findings were reported include Tannen et al. (2007, 2009), in which the results of the observational studies were fairly similar to those reported based on the RCTs. Accordingly, such findings were used to promote the use of real-world data to address important health care problems, ranging from drug safety (Mangano et al., 2006) to establishing comparative effectiveness (Pearson et al., 2011).

On the other hand, there have been situations where the results from observational studies were inconclusive or inconsistent with well-established evidence (Grodstein et al., 1996). In a series of studies, Madigan et al. (2013a,b) demonstrated that results from observational studies could be sensitive to methodological as well as database choices. In a related study, it was shown that p-values used in the reporting of nonrandomized studies are biased without proper calibration (Schuemie et al., 2014).

There is also considerable evidence pertaining to the potential of real-world data to enhance medical research (Prokosch and Ganslandt, 2009), improve efficiency in the design and conduct of clinical trials (Willke et al., 2007), and detect rare events (Brownstein et al., 2007). Although there is still some uncertainty about the role of observational studies in the drug approval process, there is optimism regarding the potential use of natural history data as a historical comparator, especially in rare disease research (FDA, 2015).

3.7 Concluding Remarks

In this chapter, we considered some of the issues associated with the use of observational studies in evidence-based medicine, and highlighted steps that need to be taken to maximize the evidentiary value in promoting public health and advancing research in medical science. Despite the proliferation of new analytical techniques to minimize bias, there is a correspondingly growing body of literature concerning the limitations of the available procedures (Madigan et al., 2013b). As in the case of RCTs, one approach to ensuring the reliability of the evidence is adherence to core statistical principles (Rubin, 2007). This of course involves entailing prospective specification of objectives and analytical procedures, guarding against the problem of multiplicity and confounding, and assuring the robustness of results through cogent sensitivity analyses. In addition, to document this, it may be necessary for such studies to be preregistered on a public website, as has become the norm for RCTs. If such steps are not taken, it is prudent for health policy decision-makers to consider the results of a particular study as being hypothesis-generating, rather than hypothesis-testing. Moreover, decision-makers should look for replication of results across multiple data sources using a variety of analytic approaches. Only by creating similar expectations for observational studies, as those that are currently applied to RCTs, as well as by assessing the rigor with which statistical analyses dealt with problems of confounding, missing data, and other issues, will observational studies become more widely used to support health care policy decisions.

References

Alemayehu, D. and J. Mardekian. 2011. Infrastructure requirements for secondary data sources in comparative effectiveness research. *J Manag Care Pharm* 17:S16–S21.

Andersen, L.W., Tobias, K., Maureen, C. et al. 2016. Early administration of epinephrine (adrenaline) in patients with cardiac arrest with initial shockable rhythm in hospital: Propensity score matched analysis. *BMJ* 353:i15773.

Austin, P.C. 2007. The performance of different propensity score methods for estimating marginal odds ratios. *Stat Med* 26:3078–3094.

Austin, P.C. 2009a. Some methods of propensity-score matching had superior performance to others: Results of an empirical investigation and Monte Carlo simulations. *Biom J* 51:171–184.

Austin, P.C. 2009b. Balance diagnostics for comparing the distribution of baseline covariates between treatment groups in propensity-score matched samples. *Stat Med* 28:3083–3107.

Austin, P.C. 2009c. Type I error rates, coverage of confidence intervals, and variance estimation in propensity-score matched analyses. *Int J Biostat* 5:Article 13. doi:10.2202/1557-4679.1146.

Austin, P.C., Grootendorst, P., and G.M. Anderson. 2007b. A comparison of the ability of different propensity score models to balance measured variables between treated and untreated subjects: A Monte Carlo study. *Stat Med* 26:734–753.

Austin, P.C., Grootendorst, P., Normand, S.L.T. et al. 2007a. Conditioning on the propensity score can result in biased estimation of common measures of treatment effect: A Monte Carlo study. *Stat Med* 26:754–768.

Berger, M.L., Mamdani, M., Atkins, D. et al. 2009. Good research practices for comparative effectiveness research: Defining, reporting and interpreting nonrandomized studies of treatment effects using secondary data sources: The International Society for Pharmacoeconomics and Outcomes Research Good Research Practices for Retrospective Database Analysis Task Force Report – Part 1. *Value Health* 12:1044–1052.

Berger, M.L, Martin, B.C., Husereau, D. et al. 2014. A questionnaire to assess the relevance and credibility of observational studies to inform health care decision making: An ISPOR-AMCP-NPC Good Practice Task Force Report. *Value Health* 17:143–156.

Bollen, K.A. and J. Pearl. 2013. Eight myths about causality and structural equation models. In *Handbook of Causal Analysis for Social Research*, ed. S.L. Morgan, pp. 301–328. Dordrecht: Springer.

Bosco, J.L., Silliman, R.A., Thwin, S.S. et al. 2010. A most stubborn bias: No adjustment method fully resolves confounding by indication in observational studies. *J Clin Epidemiol* 63:64–74.

Brookhart, M.A., Wang, P.S., Solomon, D.H. et al. 2006. Evaluating short-term drug effects in claims databases using physician-specific prescribing preferences as an instrumental variable. *Epidemiol* 17:268–275.

Brownstein, J.S., Sordo, M., Kohane, I.S. et al. 2007. The tell-tale heart: Population-based surveillance reveals an association of rofecoxib and celecoxib with myocardial infarction. *PLOS ONE* 2:e840.

CDISC. 2017. Healthcare Link Initiative. https://www.cdisc.org/standards/health-care-link (accessed September 4, 2017).

Cochran, W.G. 1968. The effectiveness of adjustment by subclassification in removing bias in observational studies. *Biometrics* 24:295–313.

Concato, J. and R.I. Horwitz. 2004. Beyond randomized versus observational studies. *Lancet* 363:1660–1661.

Cox, E., Martin, B.C., Van Staa, T. et al. 2009. Good research practices for comparative effectiveness research: Approaches to mitigate bias and confounding in the design of nonrandomized studies of treatment effects using secondary data sources: The International Society for Pharmacoeconomics and Outcomes Research Good Research Practices for Retrospective Database Analysis Task Force Report – Part II. *Value Health* 12:1053–1061.

D'Agostino, R.B. 1998. Propensity score methods for bias reduction in the comparison of a treatment to a nonrandomized control group. *Stat Med* 17:2265–2281.

Dawid, A.P. 2000. Causal thinking without counterfactuals. *J Am Stat Assoc* 95:407–424.

Deeks, J.J., Dinnes, J., D'Amico, R. et al. 2003. Evaluating non-randomised intervention studies. *Health Technol Assess.* 7:iii-x, 1–173.

Dreyer, N.A., Schneeweiss, S., McNeil, B.J. et al. 2010. GRACE principles: Recognizing high-quality observational studies of comparative effectiveness. *Am J Manag Care* 16:467–471.

Duchi, J.C., Jordan, M.I., and M.J. Wainwright. 2013. Local privacy and statistical minimax rates. *arXiv preprint arXiv:1302.3203.* https://arxiv.org/abs/1302.3203 (accessed May 10, 2017).

Food and Drug Administration (FDA). 2015. *Rare disease: Common issues in drug development—Guidance for industry.* http://www.fda.gov/downloads/Drugs/GuidanceComplianceRegulatoryInformation/Guidances/UCM458485.pdf (accessed July 1, 2016).

Greenland, S. 2005. Multiple bias modelling. *J R Stat Soc Series A.* 168:267–306.

Grodstein, F., Stampfer, M.J., Manson, J.E. et al. 1996. Postmenopausal estrogen and progestin use and the risk of cardiovascular disease. *N Engl J Med* 335:453–461.

Gu, X.S. and P.R. Rosenbaum. 1993. Comparison of multivariate matching methods: Structures, distances, and algorithms. *J Comput Graph Stat* 2:405–420.

Hastie, T., Tibshirani, R., and J. Friedman. 2009. *The Elements of Statistical Learning: Data Mining, Inference and Prediction,* 2nd edition. New York, NY: Springer.

Heckman, J.J. 2005. The scientific model of causality. *Sociol Methodol* 35:1–97.

Imbens, G.W. 2004. Nonparametric estimation of average treatment effects under exogeneity: A review. *Rev Econ Stat* 86:4–29.

International Society of Pharmacoepidemiology. 2008. Guidelines for good pharmacoepidemiology practices (GPP). *Pharmacoepidemiol Drug Saf* 17:200–208.

Johnson, M.L., Crown, W., Martin, B.C. et al. 2009. Good research practices for comparative effectiveness research: Analytic methods to improve causal inference from of nonrandomized studies of treatment effects using secondary data sources: The International Society for Pharmacoeconomics and Outcomes Research Good Research Practices for Retrospective Database Analysis Task Force Report – Part III. *Value Health* 12:1062–1073.

Kleinbaum, D.G., Sullivan, K.M., and N.D. Barker. 2013. *ActiveEpi Companion Textbook: A Supplement for Use with the ActivEpi CD-ROM,* 2nd edition. New York, NY: Springer.

Lee, B.K., Lessler, J., and E.A. Stuart. 2010. Improving propensity score weighting using machine learning. *Stat Med* 29:337–346.

Lunceford, J. K., and M. Davidian, M. 2004. Stratification and weighting via the propensity score in estimation of causal treatment effects: A comparative study. *Stat Med* 23:2937–2960.

Madigan, D., Ryan, P.B., and M.J. Schuemie. 2013b. Does design matter? Systematic evaluation of the impact of analytical choices on effect estimates in observational studies. *Ther Adv Drug Saf* 4:53–62.

Madigan, D., Ryan, P.B., Schuemie, M.J. et al. 2013a. Evaluating the impact of database heterogeneity on observational study results. *Am J Epidemiol* 15:645–651.

Madigan, D., Schuemie, M.J., and P. Ryan. 2013c. Empirical performance of the case–control method: Lessons for developing a risk identification and analysis system. *Drug Saf* 36:73–82.

Mangano, D.T., Tudor, I.C., and C. Dietzel. 2006. The risk associated with aprotinin in cardiac surgery. *N Engl J Med* 354:353–365.

Marino, S.R., Lin, S., Maiers, M. et al. 2012. Identification by random forest method of HLA class I amino acid substitutions associated with lower survival at day 100 in unrelated donor hematopoietic cell transplantation. *Bone Marrow Transplant* 47:217–226. http://www.nature.com/bmt/journal/v47/n2/full/bmt201156a.html - aff3

McCallum, R.C. and J.T. Austin. 2000. Applications of structural equation modeling in psychological research. *Annu Rev Psychol* 51:201–226.

McClellan, M., McNeil, B.J., and J.P. Newhouse. 1994. Does more intensive treatment of acute myocardial infarction in the elderly reduce mortality? Analysis using instrumental variables. *JAMA* 272:859–866.

Newhouse, J.P. and M. McClellan. 1998. Econometrics in outcomes research: The use of instrumental variables. *Annu Rev Public Health* 19:17–34.

Pearson, J.F., Brownstein, C.A., and J.S. Brownstein. 2011. Potential for electronic health records and online social networking to redefine medical research. *Clin Chem* 57:196–204.

Porter, J., Love, D., Costello, A. et al. 2015. *All-payer claims database development manual: Establishing a foundation for health care transparency and informed decision making*, APCD Council and West Health Policy Center. http://www.apcdcouncil.org/manual (accessed July 1, 2016).

Prokosch, H.U. and T. Ganslandt. 2009. Perspectives for medical informatics. Reusing the electronic medical record for clinical research. *Methods Inf Med* 48:38–44.

Psaty, B.M. and D.S. Siscovick. 2010. Minimizing bias due to confounding by indication in comparative effectiveness research: The importance of restriction. *JAMA* 304:897–898.

Robins, J.M., Blevins, D., Ritter, G. et al. 1992. G-estimation of the effect of prophylaxis therapy for pneumocystis carinii pneumonia on the survival of AIDS patients. *Epidemiol* 3:319–336.

Robins, J.M., Hernan, M.A., and B. Brumback. 2000. Marginal structural models and causal inference in Epidemiology. *Epidemiol* 11:550–560.

Rodrigues, J.J.P.C., de la Torre, I., Fernández, G. et al. 2013. Analysis of the security and privacy requirements of cloud-based electronic health records systems. *J Med Internet Res* 15:e186.

Rosenbaum, P.R. 2004. Design sensitivity in observational studies. *Biometrika* 91:153–164.

Rosenbaum, P.R. and D.B. Rubin. 1983. The central role of the propensity score in observational studies for causal effects. *Biometrika* 70:41–55.

Rothman, K.J. 1976. Causes. *Am J Epidemiol* 104:587–592.

Rubin, D.B. 2007. The design versus the analysis of observational studies for causal effects: Parallels with the design of randomized trials. *Stat Med* 26:20–36.

Schneeweiss, S. and J. Avorn. 2005. A review of uses of health care utilization databases for epidemiologic research on therapeutics. *J Clin Epidemiol* 58:323–337.

Schneeweiss, S., Rassen, J.A., Glynn, R.J. et al. 2009. High-dimensional propensity score adjustment in studies of treatment effects using health care claims data. *Epidemiol* 20:512–522.

Schuemie, M.J., Ryan, P.B., Dumouchel, W. et al. 2014. Interpreting observational studies: Why empirical calibration is needed to correct p-values. *Stat Med* 33:209–218.

Setoguchi, S., Schneeweiss, S., Brookhart, M.A. et al. 2008. Evaluating uses of data mining techniques in propensity score estimation: A simulation study. *Pharmacoepedimiol Drug Saf* 17:546–555.

Stuart, E.A. 2010. Matching methods for causal inference: A review and a look forward. *Stat Sci* 25:1–21.

Sturmer, T., Joshi, M., Glynn, R.J. et al. 2006. A review of the application of propensity score methods yielded increasing use, advantages in specific settings, but not substantially different estimates compared with conventional multivariable methods. *J Clin Epidemiol* 59:437–447.

Stukel, T.A. Fisher, E.S., Wennberg, D.E. et al. 2007. Analysis of observational studies in the presence of treatment selection bias. *JAMA* 297:278–285.

Sutter, S. 2016. Real-world evidence may find a home on breakthrough pathway. *Pink Sheet* 78/No. 26, June 27, 2016. https://www.focr.org/news/pink-sheet-real-world-evidence-may-find-home-breakthrough-pathway (accessed May 10, 2017).

Tannen, R.L., Weiner, M.G., and D. Xie. 2009. Use of primary care electronic medical record database in drug efficacy research on cardiovascular outcomes: Comparison of database and randomised controlled trial findings. *BMJ* 338:b81.

Tannen, R.L., Weiner, M.G., Xie, D. et al. 2007. A simulation using data from a primary care practice database closely replicated the women's health initiative trial. *J Clin Epidemiol* 60:686–695.

Tooth, L., Ware, R., Bain, C. et al. 2005. Quality of reporting of observational longitudinal research. *Am J Epidemiol* 16:280–288.

Uematsu, H., Kunisawa, S., Yamashita, K. et al. 2015. The impact of patient profiles and procedures on hospitalization costs through length of stay in community-acquired pneumonia patients based on a Japanese administrative database. *PLOS ONE* 10(4):e0125284. http://doi.org/10.1371/journal.pone.0125284

Velentgas, P., Dreyer, N.A., Nourjah. P. et al., eds. January 2013. *Developing a Protocol for Observational Comparative Effectiveness Research: A User's Guide*. AHRQ Publication No. 12(13)-EHC099. Rockville, MD: Agency for Healthcare Research and Quality. www.effectivehealthcare.ahrq.gov/Methods-OCER.cfm (accessed May 10, 2017).

von Elm E., Altman, D.G., Egger, M. et al. 2008. The Strengthening the Reporting of Observational Studies in Epidemiology (STROBE) statement: Guidelines for reporting observational studies. *J Clin Epidemiol* 61:344–349.

Wainwright, M.J. 2014. Constrained forms of statistical minimax: Computation, communication and privacy. https://people.eecs.berkeley.edu/~wainwrig/Barcelona14/Wainwright_ICM14.pdf (accessed on September 14, 2016).

Waning, B. and M. Montagne. 2001. *Pharmacoepidemiology Principles and Practice*. New York, NY: McGraw Hill.

Willke, R.A., Berg, R.L., Peissig, P. et al. 2007. Use of an electronic medical record for the identification of research subjects with diabetes mellitus. *Clin Med Res* 5:1–7.

Wright, J.D., Huang, Y., Burke, W.M. et al. 2016. Influence of lymphadenectomy on survival for early-stage endometrial cancer. *Obstet Gynecol* 127:109–118.

Yao, X., Abraham, N.S., Sangaralingham, L.R. et al. 2016. Effectiveness and safety of dabigatran, rivaroxaban, and apixaban versus warfarin in nonvalvular atrial fibrillation. *J Am Heart Assoc* 5:e003725. doi:10.1161/JAHA.116.003725. https://www.ncbi.nlm.nih.gov/pmc/articles/PMC4937291/ (accessed May 10, 2017).

4

Predictive Modeling in HEOR

Birol Emir, David C. Gruben, Helen T. Bhattacharyya,
Arlene L. Reisman, and Javier Cabrera

CONTENTS

4.1   Introduction 69
4.2   Exploratory Analysis and Premodeling Strategies 70
4.3   Linear Predictive Models 72
4.4   Nonlinear Predictive Models 74
4.5   Tree-Based Methods 76
         4.5.1   Ensemble Methods 76
4.6   Analyzing High-Dimensional Data 77
4.7   Other Nonparametric Regression Models 78
4.8   Software 78
4.9   Applications of Predictive Models in HEOR 79
4.10 Concluding Remarks 80
References 81

4.1 Introduction

Predictive modeling is the art and science of crafting a model that can be used to make the most accurate prediction possible (Hastie et al., 2002; Kuhn and Johnson, 2013). Traditionally statistical methods were developed mostly for the purpose of performing "inference." In the first half of the 20th century, Fisher (1925a,b) developed methodologies for statistical inference using the maximum likelihood principle. Subsequently, Neyman and Pearson (1928, 1933, 1936) formalized a theoretical framework for hypothesis testing. Savage (1954) laid the foundation for Bayesian data analysis. Afterwards, Tukey (1962) highlighted the dichotomy between statistical theory and application, and proposed a roadmap for the future of data analysis based on extensive computations and simulations. Cox (1972) extended inference and hypothesis testing methods to handle censored data. Efron (1977) introduced the concept of resampling and bootstrap methods.

More recently, Breiman et al. (1979, 1984; Breiman, 2001a,b) pioneered modern predictive modeling, which is highly dependent on computer algorithms. The approach, which draws from Tukey's ideas and Efron's resampling theory, involves classification and regression trees and the idea of ensemble prediction with random forests. Shmueli (2010) presented a very clear distinction of predictive versus inferential models, and noted the practical implications of the distinction in the modeling process. This chapter will focus on predictive modeling, with a special emphasis on health economics and outcomes research (HEOR) applications.

In particular, Section 4.1 deals with topics related to predictive modeling in general, including preprocessing, dimension reduction, and cross-validation. Section 4.2 introduces commonly used linear predictive models, particularly regression models, partial least squares, linear discriminant analysis, and penalized regression methods. Section 4.3 discusses nonlinear models, including k-nearest neighbors (k-NN), neural networks, and support vector machines. In Section 4.4, tree-based models are examined, with an emphasis placed on recursive partitioning and ensemble methods such as bagging, random forest (RF), and boosting.

Section 4.5 introduces enriched methods used to handle high-dimensional data, and a brief summary is provided in Section 4.6 for selected nonparametric regression techniques, including multivariate adaptive regression splines (MARS), projection pursuit regression, and wavelets. A list of commercially or freely available software packages is reviewed in Section 4.7. Selected examples of applications of predictive modeling techniques in HEOR are given in Section 4.8. Finally, in Section 4.9, a high-level summary of the common procedures is provided to guide the selection of methods under different scenarios.

4.2 Exploratory Analysis and Premodeling Strategies

In any modeling exercise, the data analyst needs to pay special attention to the assumptions of the models through exploratory data analysis or visualizations. This critical step includes searching for clusters, outliers, skewness, and other nonlinear structures. If the model assumptions are not validated, then appropriate remedial measures should be taken, which may include transformations or alternative model formulation. Another useful step is dimension reduction, using such techniques as principal components or factor analysis, which sometimes provides informative summarization of the variables.

Once the preprocessing step is satisfactorily executed, the analyst may proceed with the modeling step. Ignoring the initial preprocessing steps may impact the validity of the model and the accuracy of the predictions (Hoaglin et al., 1983; Cabrera and McDougall, 2002).

Cross-validation (CV) was proposed as a method for evaluating the performance of a model (Mosteller and Tukey, 1968; Lachenbruch and Mickey, 1968). The most common approach is the so-called k-fold CV, in which data are randomly divided into k groups. For the set of groups, known as a random partition, k is typically chosen to be 5, 10 or n (i.e., "leave-one-out" CV). The ingenuity of the proposal is that the ith group is predicted by fitting the model using the remaining $(k-1)$ groups. The mean square of the cross-validated residuals (CVMSR) gives an estimate of the mean squared error (MSE) that is preferred to the usual mean squared residuals from the least squares fit. The main advantage of CV is that it reduces overfitting. In the case of estimating the model parameters, we first select the models by minimizing the CVMSR over all models, or a subset of models, and obtain the final estimates of model parameters by least squares fit.

As mentioned previously, CVMSR is used to assess the performance of the selected model. For example, in all-subsets model regression in regression, one minimizes the CVMSR and chooses the model that achieves that minimum. This is an improvement over the traditional methodology that uses the Mallow's C_p criterion (Mallows, 1973). It is noted that CV does not completely eliminate overfitting, and that the CVMSR has some variability depending on the partition that is chosen. One way to reduce the variability is to average the CVMSE estimate over multiple random partitions, usually 5–20.

Bishop (1995) proposed splitting data randomly into training and test sets. The training data set is used to perform model selection using CV, and to estimate the parameters of the final model. The test data, on the other hand, is considered to be new or yet-unobserved data. The test data is used only to obtain an honest estimate of the selected model performance. The main advantage of splitting the data is that in theory, the estimate of model performance does not use data that were used to train the model, therefore providing honest estimates of model performance. However, as with CV, the performance estimate will vary depending on the choice of training and test datasets. One possibility is to repeat the random split several times, train the model on each training set, and average the performance measure (e.g., MSE) over the test datasets. It is recommended that a random mechanism that is reproducible be used to avoid any potential bias associated with the partitioning of the data into training and test sets.

More parsimonious or heavily over-parametrized models may yield small prediction errors on the training data, but are unlikely to produce similar performance on the test data. Slight changes in new test data may affect the performance of the model disproportionately. This effect is due to model overfitting on the training data. As showed in Equation 4.1, taken from Hastie et al. (2009), the MSE can be partitioned into three terms: response (error) variance (which cannot be reduced further), bias term (due to specified model), and estimation variance, which are shown respectively as follows:

$$MSE = E\left\{Y - \hat{f}(x)\right\}^2 = E\left\{Y - f(x)\right\}^2 + \left[E(\hat{f}(x)) - f(x)\right]^2 + E\left\{\hat{f}(x)) - E(\hat{f}(x))\right\}^2 \quad (4.1)$$

Both the bias and the estimation variance (the second and third terms in Equation 4.1) could be traded to reduce the MSE as much as possible. In terms of Equation 4.1, noisy data means that the variance of the outcome Y (the first term in Equation 4.1) dominates the MSE and, therefore, fancy nonlinear predictive models used in machine learning will likely not produce much improvement over the simpler linear models.

4.3 Linear Predictive Models

One of the most commonly used analytical tools is the linear regression model. Linear regression is popular because it is theoretically sound, easily implemented, and appropriate in most applications. Linear models, which express the outcome as a linear function of the regression coefficients, can be estimated using the least squares method, which under certain assumptions provides the best linear unbiased estimators. Computationally, readily available software exists that implements fast and accurate algorithms. Finally, the procedure is appropriate for many applications because most observed data are noisy and underlying relationships are not easily discerned. Hence, we are left with the choice of a first-order approximation given by a linear model.

Linear models, however, tend to have issues related to overfitting and computational instability in the presence of many predictors, especially in cases in which the predictors are highly correlated (collinear), or when the number of predictors is large relative to the number of observations, as is the case in genomics or chemistry data. For these reasons, novel methods have been introduced, including penalized regression techniques and dimension reduction approaches, which include principal component analysis (PCA) and partial least squares (PLS) (Tibshirani, 1996).

When the number of predictors is greater or equal to the number of variables, one can fit in an infinite number of hyperplanes that pass exactly by the data points. Therefore, least squares regression fails to give a unique solution. Ridge regression (Marquardt and Snee, 1975) was introduced to reduce the collinearity between the predictors by constraining the so-called squared norm of the least squares estimate ($\Sigma\beta^2 \leq c$, where the summation occurs over the predictors and c is generally chosen using CV). This constraint, moreover, can be transformed into a penalty (Lagrange multiplier) to the residual sum of squares target criteria, and optimized using standard algorithms. The resulting estimator, also called ridge regression, shrinks estimates toward the origin. Tibshirani (1996) realized that by replacing the squared norm with the absolute value norm ($\Sigma|\beta| \leq c$), an estimator may be obtained, which, besides shrinking the estimates of coefficients toward the origin, also forces the estimates of some coefficients to be exactly zero, thereby effectively performing variable selection. Accordingly, the method is dubbed least absolute shrinkage and selection operator (LASSO).

In Figure 4.1, the coefficient estimates from the least squares line (the right-most dot) are shrunk toward the origin by both LASSO (the bottom, left-most dot) and ridge (the top, left-most dot); however, LASSO forces the second coefficient (Beta2) to zero.

Zou and Hastie (2005) proposed a penalty that is a linear combination of ridge regression and LASSO penalties, called the elastic net. Finally, Friedman et al. (2008) proposed a very efficient coordinate descent algorithm to estimate the elastic net penalties, called GLMNET. This algorithm has become the standard way to compute these methods. In general, GLMNET will not exactly match the elastic net solution, but it is still a rather fast and good approximation. If we were to add elastic net constraints to Figure 4.1, it would be a curve between the LASSO (rectangle) and the ridge (larger oval) shapes.

For situations in which the number of variables is greater or equal to the number of observations, or in the presence of collinearity, an earlier attempt to address using the linear model was to use dimension reduction techniques, such as PCA or the related approach of PLS regression. Introduced by Wold (1966), PLS finds the principal components of the data in the direction of change of the response and fits a model with the first few PLS components

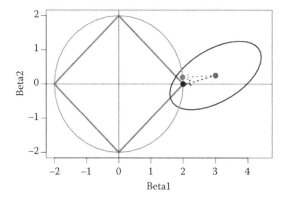

FIGURE 4.1
LASSO and ridge constraints for a linear model with two predictors, with coefficients Beta1 and Beta2 (ridge regression constraint is represented by larger oval, LASSO by the rectangle, and least squares estimation by the smaller oval).

to the response. However, it does not necessarily perform variable selection, because each PLS component is a linear combination of all of the predictors.

When the response is not continuous but instead is binary or categorical (also known as a classification problem), the same modeling strategies can be applied using logistic regression with LASSO or elastic net or logistic regression with PLS (Amaratunga et al., 2014). If we assume that the predictor variables are random (i.e., not fixed as in least squares regression), linear discriminant analysis (LDA) provides a linear partition of the predictor space by the response classes. LDA, suggested by Fisher (1936), assumes that the predictors follow a mixture of multivariate normal distributions with a common covariance matrix and different group mean vectors. The objective is to estimate the parameters of the mixture of normal distributions. Each data point is assigned to the class corresponding to the largest value of the density functions at that point.

In some cases, it is possible to find relationships between the predictors and the outcome that are not linear in the predictors (e.g., they can be quadratic or polynomial), while the outcome is expressed as a linear function of the regression coefficients. In other cases, the relationships may be more complex and may require special nonlinear models to be fit. In the next section, advanced predictive models that are nonlinear, including neural networks and support vector machines, are addressed.

4.4 Nonlinear Predictive Models

There are two situations involving the fitting of a nonlinear model. In the first case, there is an explicitly known nonlinear function that describes the relationship between the predictors and the response. For example, in fitting dose-response curves, it is customary to use an *EMAX* model (Bates and Watts, 1988). A four-parameter *EMAX* model is given in Equation 4.2:

$$g(D) = E(Y \mid D) = E_0 + \frac{E_{\max}D^\lambda}{ED_{50}^\lambda + D^\lambda} \tag{4.2}$$

where E_0 is the expected response at baseline (absence of dose), $E_{\max}$ is the asymptotic maximum dose effect, ED_{50} is the dose that produces 50% of the maximal effect, and λ is sometimes called the Hill or slope parameter. When the form of the relationship is known, the unknown parameters may be estimated using conventional numerical methods.

The second case arises when the nonlinear relationship is unknown. This case is very common with high-dimensional data, where relationships are usually little understood. In what follows, the focus is on this latter situation.

k-NN is one of the earliest nonlinear algorithms proposed to predict or classify a new data point (Fisher, 1936). The algorithm seeks the *k* nearest points, among the observations, to the new data point, with respect to a Euclidian or some other specified distance measure. In order to classify a new data point, a majority rule is applied. The number *k* of nearest neighbors is a tuning parameter, and can be optimized by minimizing the MSE or classification error over a range of values of *k*. For predicting a continuous response, we can either use an average or a linear prediction of the *k*-NN responses. The resulting fit is a nonlinear function because the fit always changes when the neighborhood changes. Another important use of the *k*-NN is for imputing missing values, by replacing the missing value with the average of the *k*-NN values of that predictor (Dixon, 1979).

Neural networks were inspired from the perceptron learning rule of Rosenblatt (1958), and were formally introduced by Rumelhart et al. (1986). The idea relies on a structure of one or more hidden layers of nodes between the predictors and the outcome of interest. The predictors are used as inputs for the nodes of the first hidden layer where, at each node, a nonlinear function (e.g., sigmoidal) is applied to a linear combination of the predictors to produce an output. The outputs of the nodes in a hidden layer are in turn used as inputs to the nodes in the next layer, and so on, until we arrive at the last layer of the neural net. The outputs of the last layer are the linear predictors of the response. The complexity of this model is that the algorithm will evaluate a sequence of functions of functions, which makes the procedure computationally intensive, difficult to fit, and almost impossible to interpret. Since the number of nodes and hidden layers may be quite large, the neural net may be heavily parameterized and, therefore, likely to overfit. To reduce the overfitting, one may use CV.

Neural nets were highly popular in the late 1990s and early 2000s, but their importance gradually dissipated due to the concerns raised in the previous paragraph (computationally intensive, difficult to fit, almost impossible to interpret). However, lately, the use of neural nets with many hidden layers—also called deep learning—have become popular again in the machine learning literature (Goodfellow et al., 2016). The renewed interest arrives in combination with the enormous parallel computing resources that are becoming more and more available with the appearance of big hubs such as Amazon or Google. Nonetheless, it is still unclear whether the incremental cost is worth the benefits, as compared with cheaper alternatives, like support vector machines (SVMs).

Vapnik (1995) proposed a SVM as a classification method. SVMs take a very different approach than Fisher's discriminant analysis or logistic regression in that they optimize some probabilistic model in *all* the data. A SVM focuses only in finding hyperplanes that locally optimize the separation between the classes in the region in which they overlap. The approach uses optimization with constraints to estimate the hyperplane going through boundary points, called support vectors, that produces the best separation. There is also an alternative SVM algorithm for the regression problem (Gunn, 1998).

4.5 Tree-Based Methods

Classification and regression trees have been the focus of research in the statistical literature (Breiman et al., 1984; Hartigan, 1975; Hawkins and Kass, 1982; Sonquist and Morgan, 1964). In recent years, they have also become a popular topic in computer science and machine learning (Quinlan, 1993).

Tree-based methods involve splitting the data into mostly two buckets and achieving minimization using suitably chosen optimization criteria. The standard algorithm proposed in Breiman et al. (1984) grows the tree in a recursive fashion until the buckets are "small" enough. Tree growth is followed by a pruning step to eliminate those splits that don't pass a pruning criterion, such as CV or Mallows' C_p (Mallows, 1973). The purpose of the pruning step is to avoid overfitting.

The main advantage of the use of a classification tree is that the results are readily interpretable and, hence, easy to communicate to practitioners. Further, the approach requires minimal preprocessing and is reasonably fast to compute even with big data, requiring a computation time only of order nxp (the number of patients times the number of predictors). In linear modeling, trees are also useful for detecting deep interactions among the predictors.

The main disadvantage of a tree is that it is a step-function approximation to the response surface along the predictors' axes. This could result in a very poor approximation to the true response surface, which is more likely to be continuous, or at least not a step function. As a result, a tree-based approach may not be as powerful as linear or nonlinear models, when the underlying assumptions for these models hold.

4.5.1 Ensemble Methods

Breiman (1996) introduced the idea of ensemble methods to improve upon the predictive ability of learning algorithms. The approach involves taking a bootstrap sample, called an *in-bag* sample, from the training set. If the bootstrap sample is equal to the size of the training set, on average, then the in-bag sample will have 63.2% of the original data. The remaining 36.8% of observations form the *out-of-bag* sample. This process is repeated many times, and is referred to as *bagging* or bootstrap aggregation. The in-bag samples are used to train the model, while the out-of-bag samples will help with evaluating the accuracy of prediction. The final prediction is accomplished either by a majority vote for classification or, alternatively, taking an average over all of the out-of-bag predictions for regression. The algorithm is fairly general and could be applied to any learning method, including trees, LASSO, or SVM.

Breiman (2001a) argued that for classification and regression trees, *bagging* could be improved, and proposed a related algorithm, RF. In this approach, in addition to bagging, the tree algorithm is modified so that at each node k, p predictors are selected at random to perform the best split. Usually k is a fixed number, often set equal to $\sqrt{p}$ or $p/3$, or chosen by CV.

Boosting is an ensemble method that builds a sequence of models that are not independent (Schapire, 1990). Each model uses the same dataset with a different set of weights. Specifically, if an observation is misclassified or has a high residual, then the weight for that observation is increased for the next model iteration. If the prediction is reasonably good, then the weights of the observations are decreased. The boosting prediction is an aggregate, such as the mean or the majority vote of the predictions over the entire sequence of models.

4.6 Analyzing High-Dimensional Data

When dealing with high-dimensional data, such as genomics data, one should recognize that the high dimensionality could easily hide the signal in the data in favor of randomly occurring spurious signals. One remedy is to use penalized methods such as LASSO; however, such methods would not generally uncover signals of dimension greater than $n/\log p$ (Cai and Zhang, 2016). For example, in a biological experiment with 100 samples and 100,000 predictors, LASSO would likely work for signals of eight or fewer predictors. Therefore, the penalized methods would not work unless they are preceded by some dimension reduction.

Another alternative approach is to use ensemble methods, such as RF. In the above biological experiment, suppose we have 100,000 predictors containing a signal of eight predictors plus noise. If we select 300 predictors at random, then the probability of including one or more of the eight predictors is only 2.4%. For a tree with 10 nodes, the chance that all eight predictors are excluded from the tree is 78.6%. So, it can be argued that RF is not a good alternative in this case either.

In practice, the analyst preprocesses the data by discarding most of the predictors using a simple univariate threshold that relates to the correlation between the outcome and the predictors. Nevertheless, the selection of this threshold is often unsatisfactory, because different thresholds may yield different predictions.

Amaratunga et al. (2008) introduced the idea of enriching to address the above problem. The approach involves assigning a weight to each variable, as opposed to each observation. The weight incorporates the strength of the univariate relationship between the outcome and the predictor, as corrected for false discoveries (Benjamini and Hochberg, 1995). The enriched version of RF may be extended to LASSO, PCA, and PLS.

4.7 Other Nonparametric Regression Models

There are several other alternative methods for nonparametric multivariate regression. Typical examples include projection pursuit regression (PPR), MARS, and wavelets.

PPR, proposed by Friedman and Stuetzle (1981), is a nonparametric approach that approximates the regression surface iteratively as a sum of smooth functions of linear combinations of the predictors. MARS, introduced by Friedman (1991), is also a nonparametric regression technique, but involves a weighted sum of special functions (basis functions) to approximate a smooth function.

Wavelets are a special type of basis functions that are relatively easy to compute and give spatially adaptive estimators—that is, they permit accommodation of local features in the data. Wavelet regression is especially successful when the function is discontinuous, when it is continuous but has sharp changes or discontinuities in the derivative or, more generally, when the function is not smooth (Daubechies, 1990, 1995).

4.8 Software

There are many software packages to implement the techniques discussed in this chapter. In particular, R and Weka, which can be downloaded freely, provide several algorithms that can readily be applied to data. The main R packages include *glmnet, randomForest, nnet, e1071, rpart,* and *erf.* The R libraries *MASS* and *caret* support many additional tools that can be used for machine learning.

Commercially available software packages include RapidMiner, MATLAB®, SAS JMP, Knime, Orange, and SPSS Modeler. RHadoop and sparkR are developed to work under the Hadoop and Spark big data platforms, respectively. As the sizes of datasets increase, it may be essential to use larger computational resources, such as Amazon Web Services or other cloud services. These computational services are very convenient and transparent to the users who can run software such as RStudio on a webpage. Nonetheless, they are not often desirable because of data privacy and security issues.

4.9 Applications of Predictive Models in HEOR

The HEOR literature abounds with examples of applications of predictive models. Cabrera and McDougall (2002) analyzed hospital summary data related to orthopedics products from all hospitals in the USA during 1995–1996. An exploratory analysis, consisting of variable transformations followed by PCA, was used to group the variables that were input into a cluster analysis for further modeling within each cluster. Odgers et al. (2016) implemented LASSO regression to classify rheumatoid arthritis patient outcomes, using electronic medical records (EMR) and biomedical linked open data.

Use of a penalized regression approach in an acquired immune deficiency syndrome (AIDS) trial is reported in Lu et al. (2013). This study was aimed at finding optimal treatments to maximize CD4 counts in AIDS patients. The authors built a model to select optimal treatment strategies while performing variable selection using LASSO.

Emir et al. (2017) applied a spectrum of machine learning algorithms to predict treatment responders using data from an observational study. Patients with neuropathy were followed for six weeks while receiving treatment. Daily diary pain scores were recorded using a numeric rating between 0 (no pain) and 10 (worst imaginable pain). The response to treatment was defined as achieving a 50% or more reduction in the average pain scores. The paper compared the prediction accuracy of several predictive models, including k-NN, PLS, LDA, PLS, and others. The performance of k-NN was comparable to those of more sophisticated, computationally intensive algorithms, including CART, SVM, RF, and boosting. That said, boosted classification and RF models were marginally better performers for predicting treatment response.

Raghupathi and Raghupathi (2017) used neural network approaches to examine how behavioral habits (e.g., smoking, inactivity) and demographics contribute to the incidence of chronic diseases, with the goal of encouraging preventative health care. Miotto et al. (2016) applied unsupervised deep learning to predict the health states of patients using a large-scale EMR database.

Araújo et al. (2016) used SVM, k-NN, and other algorithms to learn about health care preauthorization. In order to assist dental health care professionals in decision-making, a potential decision was generated by learning from a health insurance provider's database.

Francis et al. (2011) used SVM to create an expert system to augment the manual review of medical bills, especially for the detection of fraud and abuse. Currently, manual review is completed by highly skilled workers in a time-consuming process. However, the laws and regulations that require prompt payment has incited pressure to develop automation methods to speed the review process.

Emir et al. (2015) applied RF for the purpose of predictive modeling in a situation where there was a severe class imbalance. A range of demographics and clinical and health care resource utilization variables were extracted and served as input into the model.

4.10 Concluding Remarks

This chapter provides a brief historical account of predictive modeling methods and gives heuristic explanations of the main steps involved in model building, including preprocessing, learner training, and cross-validation. Alternative algorithms for linear and nonlinear models, as well as tree-based methods, are reviewed. While the chapter's emphasis is on non-Bayesian approaches, it should be noted that Bayesian alternatives are also available to some of the methods discussed.

Table 4.1 gives a summary of relevant features of some of the methods discussed, with an emphasis on the sample size (n), number of predictors (p), and the underlying relationship structure.

As shown in Table 4.1, when $n > p$ (and model assumptions are satisfied), standard statistical predictive methods (e.g., traditional generalized linear

TABLE 4.1

A Summary of Methods

n samples p predictors	Size	Underlying Relationship		
		Linear	Nonlinear and Smooth	Nonlinear but not Smooth
$n > p$	n small	Bayesian or GLM	GLM Neural nets/deep learning/SVM Ensemble (RF, boosting)	Wavelets Trees Bagging, RF
	n large/p is *likely* large (Big Data)	GLM/penalized (LASSO) Fast algorithms (e.g., 1-step logistic regression)	Deep learning (cloud computing with many GPUs) Ensemble (RF, boosting)	
$n \leq p$ high dimensional (mega or giga variate)	$n/\log p > 100$ *n and p both large*	LASSO/elastic net PLS	Penalized SVM RF, boosting	Wavelets Trees Bagging, RF
	$n/\log p < 100$ *n small and p large*	Enriched versions of RF LASSO and elastic net	Enriched RF Enriched penalized SVM	Single tree Enriched RF

Note: GLM = general linear model; GPU = graphics processor units; PLS = partial least squares; RF = random forests; SVM = support vector machines.

models or penalized models) may be appropriate. When the underlying relationships among the predictors and the outcome are assumed to be nonlinear, then machine learning methods, including wavelets or tree-based methods, are suggested. For big data, where computational performance may be an issue, fast algorithms may need to be implemented (Genkin et al., 2007).

As stated in Table 4.1, when $n \leq p$, it may be essential to apply penalized statistical models or dimension reduction techniques, such as PLS. When the underlying relationships among the predictors and the outcome are assumed to be nonlinear, penalized SVM and RF are proposed. Enriched methods, including enriched versions of LASSO, elastic net, RF, and SVM are recommended when n is much smaller than p ($n/\log p < 100$). Finally, Table 4.1 should be taken as no more than a set of general guidelines, as each situation must be handled individually depending on the specifics that may influence the analysis path that is chosen.

References

Amaratunga, D., Cabrera, J., and Y.S. Lee. 2008. Enriched random forests. *Bioinformatics* 24:2010–2014.

Amaratunga, D., Cabrera, J., and Z. Shkedy. 2014. *Exploration and Analysis of DNA Microarray and Other High-Dimensional Data*, 2nd edition. Hoboken, NJ: John Wiley & Sons.

Araújo, F.H.D., Santana, A.M., and P. de A. Santos Neto. 2016. Using machine learning to support healthcare professionals in making preauthorisation decisions. *Int J Med Inform* 94:1–7.

Bates, D.M. and D.G. Watts. 1988. *Nonlinear Regression Analysis and its Applications*. Hoboken, NJ: John Wiley & Sons.

Benjamini, Y. and Y. Hochberg. 1995. Controlling the false discovery rate: A practical and powerful approach to multiple testing. *J R Stat Soc Series B Methodol* 57:289–300.

Bishop, C.M. 1995. *Neural Networks for Pattern Recognition*. New York, NY: Oxford University Press.

Breiman, L. 1996. Bagging predictors. *Mach Learn* 24:123–140.

Breiman, L. 2001a. Random forests. *Mach Learn* 45:5–32.

Breiman, L. 2001b. Statistical modeling: The two cultures (with comments and a rejoinder by the author). *Stat Sci* 16:199–231.

Breiman, L., Friedman, J., Stone, C.J. et al. 1984. *Classification and Regression Trees*. Boca Raton, FL: Chapman & Hall/CRC Press.

Breiman, L., Stone, C.J., and J.D. Gins. 1979. *New Methods for Estimating Tail Probabilities and Extreme Value Distributions (No. TSC-PD-A226-1)*. Santa Monica, CA: Technology Service Corporation.

Cabrera, J. and A. McDougall. 2002. *Statistical Consulting*. New York, NY: Springer.

Cai, T.T. and A. Zhang. 2016. Inference for high-dimensional differential correlation matrices. *J Multivar Anal* 143:107–126.

Cox, D.R. 1972. Regression models and life-tables. *J R Stat Soc Series B Methodol* 34:187–220.

Daubechies, I. 1990. The wavelet transform, time-frequency localization and signal analysis. *IEEE T Inform Theory* 36:961–1005.

Daubechies, I. 1995. *Ten Lectures on Wavelets, CBMS-NSF Regional Conference Series in Applied Mathematics*, vol. 61. Philadelphia, PA: Society for Industrial and Applied Mathematics.

Dixon, J. K. 1979. Pattern recognition with partly missing data. *IEEE Trans Syst Man Cybern* 9:617–621.

Efron, B. 1977. *Bootstrap methods: Another look at the jackknife - report No. 32.* Palto Alto, CA: Stanford University, Division of Biostatistics.

Emir, B., Johnson, K., Kuhn, M. et al. 2017. Predictive modeling of response to pregabalin for the treatment of neuropathic pain using 6-week observational data: A spectrum of modern analytics applications. *Clin Ther* 39:98–106.

Emir, B., Masters, E.T., Mardekian, J. et al. 2015. Identification of a potential fibromyalgia diagnosis using random forest modeling applied to electronic medical records. *J Pain Res* 8:277–288.

Fisher, R.A. 1925a. Theory of statistical estimation. *Math Proc Cambridge* 22:700–725.

Fisher, R.A. 1925b. *Statistical Methods for Research Workers*, 5th edition–revised and enlarged. Edinburgh: Oliver and Boyd.

Fisher, R.A. 1936. The use of multiple measurements in taxonomic problems. *Ann Hum Genet (formerly Annals of Eugenics)* 7:179–188.

Francis, C., Pepper, N., and H. Strong. 2011. Using support vector machines to detect medical fraud and abuse. *Proceedings of the 2011 Annual International Conference of IEEE Engineering in Medicine and Biology Society*, pp. 8291–8294.

Friedman, J. H. 1991. Multivariate adaptive regression splines. *Ann Stat* 19:1–67.

Friedman, J., Hastie, T., and R. Tibshirani. 2008. Regularization paths for generalized linear models via coordinate descent. *J Stat Softw* 33:1.

Friedman, J. H. and W. Stuetzle. 1981. Projection pursuit regression. *J Am Stat Assoc* 76:817–823.

Genkin, A., Lewis, D.D., and D. Madigan. 2007. Large-scale Bayesian logistic regression for text categorization. *Technometrics* 49:291–304.

Goodfellow, I., Bengio, Y., and A. Courville. 2016. *Deep Learning.* Cambridge, MA: MIT Press.

Gunn, S.R. 1998. Support vector machines for classification and regression. *ISIS Technical Report* 14:85–86.

Hartigan, J.A. 1975. *Clustering Algorithms.* New York, NY: John Wiley & Sons.

Hastie, T., Tibshirani, R., and J. Friedman. 2009. *The Elements of Statistical Learning: Data Mining, Inference and Prediction*, 2nd edition. New York, NY: Springer.

Hastie, T., Tibshirani, R., and J.H. Friedman. 2002. *The Elements of Statistical Learning.* New York, NY: Springer.

Hawkins, D.M. and G.V. Kass. 1982. Automatic interaction detection. In *Topics in Applied Multivariate Analysis*, ed. D.H. Hawkins, pp. 269–302, *Cambridge: Cambridge University Press.*

Hoaglin, D.C., Mosteller, F., and J.W. Tukey, eds. 1983. *Understanding Robust and Exploratory Data Analysis.* New York, NY: John Wiley & Sons.

Kuhn, M. and K. Johnson. 2013. *Applied Predictive Modeling.* New York, NY: Springer.

Lachenbruch, P.A. and M.R. Mickey. 1968. Estimation of error rates in discriminant analysis. *Technometrics* 10:1–12.

Lu, W., Zhang, H.H., and D. Zeng. 2013. Variable selection for optimal treatment decision. *Stat Methods Med Res* 22:493–504.

Mallows, C.L. 1973. Some comments on cp. *Technometrics* 15:661–675.

Marquardt, D.W. and R.D. Snee. 1975. Ridge regression in practice. *Am Stat* 29:3–20.

Miotto, R., Li, L., Kidd, B. et al. 2016. Deep patient: An unsupervised representation to predict the future of the patients from the electronic health records. Sci Rep 6, article number 26094. doi:10.1038.srep26094. https://www.nature.com/articles/srep26094 (accessed May 28, 2017).

Mosteller, F. and J.W. Tukey. 1968. Data analysis, including statistics. In *Handbook of Social Psychology*, vol. 2, 2nd edition, eds. G. Lindzey and E. Aronson, pp. 80–203. Reading, MA: Addison-Wesley.

Neyman, J. and E.S. Pearson. 1928. On the use and interpretation of certain test criteria for purposes of statistical inference: Part I. *Biometrika* 20:175–240. (Reprinted in J. Neyman and E.S. Pearson Joint Statistical Papers 1–67. Cambridge: Cambridge University Press, 1967.)

Neyman, J. and E.S. Pearson. 1933. On the problem of the most efficient tests of statistical hypotheses. *Phil Trans R Soc A* 231:289–337. (Reprinted in J. Neyman and E.S. Pearson Joint Statistical Papers 140–85. Cambridge: Cambridge University Press, 1967.)

Neyman, J. and E.S. Pearson. 1936. Contributions to the theory of testing statistical hypotheses. *Stat Res Mem* 1:1–37. (Reprinted in J. Neyman and E.S. Pearson Joint Statistical Papers 203–39. Cambridge: Cambridge University Press, 1967.)

Odgers, D.J., Tellis, N., Hall, H. et al. 2016. Using LASSO regression to predict rheumatoid arthritis treatment efficacy. *AMIA Jt Summits Sci Proc.* 2016:176–183. https://www.ncbi.nlm.nih.gov/pmc/articles/PMC5001752/ (accessed May 28, 2017).

Quinlan, R. 1993. *C4.5: Programs for Machine Learning.* San Mateo, CA: Morgan Kaufmann Publishers.

Raghupathi, V. and W. Raghupathi. 2017. Preventive healthcare: A neural network analysis of behavioral habits and chronic diseases. *Healthcare* 5:8. doi:10.3390/healthcare5010008. http://www.mdpi.com/2227-9032/5/1/8 (accessed May 28, 2017).

Rosenblatt, F. 1958. The perceptron: A probabilistic model for information storage and organization in the brain. *Psychol Rev* 65:386–408.

Rumelhart, D.E., Hinton, G.E., and R.J. Williams. 1986. Learning representations by back-propagating errors. *Nature* 323:533–536.

Savage, L.J. 1954. *The Foundations of Statistics.* New York, NY: John Wiley & sons.

Schapire, R.E. 1990. The strength of weak learnability. *Mach Learn* 5:197–227.

Shmueli, G. 2010. To explain or to predict? *Stat Sci* 25:289–310.

Sonquist, J.A. and J.N. Morgan. 1964. *A Detection of Interaction Effects: A Report on a Computer Program.* Ann Arbor, MI: Survey Research Centre Institute for Social Research, University of Michigan.

Tibshirani, R. 1996. Regression shrinkage and selection via the lasso. *J R Stat Soc Series B Methodol* 58:267–288.

Tukey, J.W. 1962. The future of data analysis. *Ann Math Stat* 33:1–67.

Vapnik, V. 1995. *The Nature of Statistical Learning Theory.* New York, NY: Springer.

Wold, H. 1966. Estimation of principal components and related models by iterative least squares. *In Multivariate Analysis*, ed. P. Krishnaiah, pp. 391–420. New York, NY: Academic Press.

Zou, H. and T. Hastie. 2005. Regularization and variable selection via the elastic net. *J Roy Stat Soc Series B Methodol* 67:301–320.

5

Methodological Issues in Health Economic Analysis

Demissie Alemayehu, Thomas Mathew, and Richard J. Willke

CONTENTS

5.1 Introduction 85
5.2 Cost-Effectiveness Criteria and Statistical Inference 86
    5.2.1 Traditional Measures of Cost-Effectiveness 87
    5.2.2 The Cost-Effectiveness Acceptability Curve (CEAC) 88
    5.2.3 Statistical Inference for Cost-Effectiveness Measures 89
    5.2.4 The Generalized Pivotal Quantity (GPQ) Approach 91
    5.2.5 An Example 93
    5.2.6 A Probabilistic Measure of Cost-Effectiveness 95
5.3 Incorporating Data from Indirect Comparisons and
    Observational Studies 98
    5.3.1 Indirect Comparisons 99
    5.3.2 Effectiveness Data from Observational Studies 101
    5.3.3 Issues with Cost Data 102
5.4 Decision Analysis 104
    5.4.1 Steps in Decision Analysis 105
    5.4.2 Outcome Measures 105
    5.4.3 Decision Trees 107
    5.4.4 Markov Models 108
    5.4.5 Use in CEA 110
    5.4.6 Sensitivity Analysis 110
    5.4.7 CEA Example Using a Markov Model 111
5.5 Conclusion 115
Acknowledgment 116
References 116

5.1 Introduction

When alternative treatment options are available, a decision regarding the allocation of scarce health care resources can be made on the basis of comparative data involving both costs and health outcomes. Cost-effectiveness analysis (CEA) is a commonly used approach, addressing a useful purpose as a combined measure of both the costs and the health outcomes associated with competing intervention strategies (Luce et al., 1996). In this framework, when data are available for a common outcome measure for the treatment options of interest, results are typically presented in terms of cost per unit of the health outcome, for example, cost per untoward event prevented, cost per life year gained (Weinstein and Stason, 1977), or cost per quality-adjusted life year (Luce et al., 1996). This feature of CEA makes it appealing as a procedure of choice in decision-making for both policy-makers and health care providers.

Despite the widespread use of CEA in medical research and health technology assessment, there are certain methodological issues that need to be addressed to ensure the appropriate application of the obtained results (Doubilet et al., 1986). Like any other measure based on sampled data, sources of uncertainty around cost-effectiveness estimates must be carefully considered and incorporated into inferences that involve those results. In particular, CEA presents some challenges in that regard due to its use of ratio measures, as well as to the mixed stochastic and assumption-based nature of health economic modeling.

In addition, since the outcome measures used to assess effectiveness in the original trials may not be similar, it may not always be feasible to incorporate all available data in evidence synthesis. Further, in certain situations, head-to-head comparative data from randomized controlled trials (RCTs), which are the gold standard for evidence-based medicine, may not be available for treatments that need to be studied, or the available data from the RCTs may not be reflective of real-world experiences.

In view of the considerable impact of CEA results on health care decision-making, there has been increased focus in the literature in terms of buttressing the reliability of the results (Drummond et al., 1997; O'Brien et al., 1994; Weinstein et al., 1996). In this chapter, we provide a general overview of the standard approaches and recent developments in CEA, and discuss issues associated with the incorporation of data from non-RCT sources. In addition, we consider the role of decision analysis in CEA, including the use of Markov models when comparing treatment options in patients who experience changes in health states over time.

5.2 Cost-Effectiveness Criteria and Statistical Inference

Extensive literature is available on the criteria to be used for CEA and the corresponding statistical inference. In this section, we shall not provide a detailed discussion of these; instead, we draw the attention of the reader to some of the relevant references. Our focus here is on some of the very recent developments.

5.2.1 Traditional Measures of Cost-Effectiveness

While evaluating treatments regarding their costs and health benefits, traditional criteria involve comparison mainly based on the means. The most common criteria are the incremental cost-effectiveness ratio (ICER) and the incremental net benefit (INB). Another criterion that is less often used in the average cost-effectiveness ratio (ACER). In what follows, we provide a brief review of these criteria in the context of data obtained from RCTs.

We shall consider the setup of a two-arm RCT, with patients being randomized to two treatments. Let μ_C^j C^j and μ_E^j E^j be random variables representing the cost and the effectiveness measures for the patients in the jth group ($j = 1, 2$). Furthermore, let μ_C^j and μ_E^j be the respective population mean cost and population mean effectiveness in the jth group. The ICER parameter is the incremental cost of achieving one unit of effectiveness from using the first treatment instead of the second, and is defined as the ratio of the difference between expected mean costs and the difference between expected mean effectiveness, as follows:

$$ICER = \frac{\mu_C^1 - \mu_C^2}{\mu_E^1 - \mu_E^2} \tag{5.1}$$

As a measure of cost-effectiveness, the ICER is the most common one used by health economists, and allows for decision-making that involves explicit tradeoffs between health gain and incremental costs due to treatment. Nevertheless, it suffers from several drawbacks, including that an insignificant difference in effectiveness leads to a near-zero denominator. Furthermore, the ICER cannot distinguish between a case in which the first treatment is more effective and less costly than the second (i.e., is "dominant") and a case in which the first treatment is less effective and more costly than the second (i.e., is "dominated"). Also, being a ratio parameter, the ICER presents further difficulties when it comes to statistical inference.

A second criterion that has been suggested in the literature, the INB, is free of these drawbacks. The definition of the INB requires the specification of a willingness-to-pay (WTP) parameter λ, which is the amount a policy-maker is willing to pay for an increase in effectiveness of one unit. The quantity INB is defined as the difference of incremental effectiveness (multiplied by λ) and the incremental cost, and we shall use the notation INB (λ) in order to make explicit the dependence of INB on λ:

$$\text{INB}(\lambda) = \lambda\left(\mu_E^1 - \mu_E^2\right) - \left(\mu_C^1 - \mu_C^2\right) \tag{5.2}$$

Proposed as an alternative of the ICER, INB (λ) is the net benefit of giving a patient treatment 1 rather than treatment 2. When INB(λ) > 0, we conclude that the increased amount the policy-makers are willing to pay exceeds the growth in cost; we conclude the opposite when INB(λ) < 0. The ICER and the INB are clearly related algebraically; INB (λ) > 0 is equivalent to ICER <λ. Statistical inference concerning ICER and INB (λ) are discussed extensively in the book by Willan and Briggs (2006).

Another parameter that has received some attention in the literature is the ACER, which is the average cost spent per unit of effectiveness. If μ_C and μ_E are, respectively, the population mean cost and mean effectiveness for a single group, then the ACER is defined as the ratio μ_C / μ_E. One can also define an "incremental ACER," say ΔACER, as follows:

$$\Delta\text{ACER} = \frac{\mu_C^1}{\mu_E^1} - \frac{\mu_C^2}{\mu_E^2} \tag{5.3}$$

The roles of the parameters ICER, ACER, and ΔACER have been compared in the literature (Bang and Zhao, 2012; Briggs and Fenn, 1997; Hershey et al., 2003; Hoch and Dewa, 2008; Laska et al., 1997a, b). The article by Laska et al. (1997a) stated that the ACERs play an important role in CEA, and that decision-makers could use either the ACER or the ICER in some scenarios. This assertion was criticized by Briggs and Fenn (1997), who argued that the ACER comparison does not provide the guidance necessary for choosing between two treatments, as opposed to the ICER. However, Laska et al. (1997b) refuted this assertion. The careful arguments presented in a later article by Hoch and Dewa (2008), and the results based on hypothetical scenarios presented to physicians reported in Hershey et al. (2003), indicate that the use of ACERs for decision-making could lead to choosing treatments with poor cost-effectiveness. We will not further elaborate on this issue here; however, the interested reader may refer to the just-cited articles for more information and insight on the comparison between the ACER and ICER.

5.2.2 The Cost-Effectiveness Acceptability Curve (CEAC)

We noted above that a positive value of INB (λ) indicates the cost-effectiveness of the first treatment over the second. However, INB (λ) depends on unknown parameters, and it also depends on the chosen value of λ. Thus, one way to assess if INB(λ) > 0 is to evaluate the probability that an estimate of the INB (λ) is positive. Such an estimate, plotted against λ, is referred to as the CEAC. If (Cost, Effectiveness) follows a bivariate normal distribution, then an unbiased estimator of INB (λ) can be obtained by replacing μ_C^1, μ_C^2, μ_E^1, and μ_E^2 by their corresponding sample means. The estimator so obtained, denoted by $\widehat{\text{INB}}(\lambda)$, is an unbiased estimator of INB (λ). Furthermore, $\widehat{\text{INB}}(\lambda)$ is normally distributed. In this case, $P[\widehat{\text{INB}}(\lambda) > 0]$ is easy to compute and the plot of the CEAC can be obtained.

Nonetheless, if bivariate normality does not hold, an exact calculation of such a probability is not possible, and one may have to resort to asymptotic normality, or bootstrap methods, in order to obtain the CEAC. We refer to Fenwick and Byford (2005) and Fenwick et al. (2001) for guidance on CEACs and their role in CEA. Estimation of CEACs is discussed in detail in Nixon et al. (2005). Various facts and fallacies associated with CEACs are addressed in Fenwick et al. (2004). A related concept is that of the cost-effectiveness acceptability frontier (CEAF), which is the probability that the treatment option with the highest net benefit is cost-effective, assuming a willingness to pay value equal to λ. We refer to Barton et al. (2008) for a discussion and comparison of the CEAC and CEAF.

5.2.3 Statistical Inference for Cost-Effectiveness Measures

The calculation of point estimates of the parameters ICER, INB (λ), ACER and ΔACER is straightforward; however, the investigation of their statistical properties is less so. Various approaches have been suggested in the literature in order to construct confidence intervals for ICER and INB (λ). The methodologies include Fieller's theorem (Briggs and Fenn, 1997; Briggs et al., 1999; Laska et al., 1997a; Willan and O'Brien, 1996), Taylor series expansion (Briggs and Fenn, 1997; Briggs et al., 1999; Chaudhary and Stearns, 1996; O'Brien et al., 1994), nonparametric methods such as various bootstrap methods (Briggs and Fenn, 1997; Briggs et al., 1999; Chaudhary and Stearns, 1996), the concept of generalized pivotal quantity (Bebu et al., 2016a), and Bayesian methods (Baio, 2012; Thompson and Nixon, 2005). The combination of Fieller's method with the bootstrap has also been proposed (Jiang et al., 2000). The parametric procedures very often assume normality or asymptotic normality.

Fieller's method is based on such a normality assumption; however, it is well known that the method may not produce a finite confidence interval. An aspect to be noted is that cost data often exhibit skewness, and log-normality appears more appropriate. Under log-normality, the inference problems of interest involve rather complicated parametric functions. In order to illustrate this, suppose that marginally, the costs follow a lognormal distribution, and the effectiveness measures follow a normal distribution; we shall refer to this as the lognormal/normal scenario. Thus in the lognormal/normal setup, $\ln(C^j)$ follows a normal distribution with mean μ_{lC}^j, and E^j follows a normal distribution with mean μ_E^j, where C^j is the cost and E^j is the effectiveness measure for the patients in the jth group ($j = 1, 2$). Clearly, μ_{lC}^j is the population mean of the log-transformed costs.

A model for the bivariate random variable (Cost, Effectiveness) can now be specified as

$$\begin{pmatrix} \ln(C^j) \\ E^j \end{pmatrix} \sim N\left(\begin{pmatrix} \mu_{IC}^j \\ \mu_E^j \end{pmatrix}, \Sigma^j \right) \tag{5.4}$$

In view of the log-normality of the C^j, the mean of C^j is given by

$$E(C^j) = \exp\left(\mu_{IC}^j + \frac{1}{2}\Sigma_{11}^j \right)$$

where Σ_{11}^j is the first diagonal element of the covariance matrix Σ^j, $j = 1, 2$. Furthermore, the ICER has the expression

$$\text{ICER} = \frac{\exp\left(\mu_{IC}^1 + \frac{1}{2}\Sigma_{11}^1 \right) - \exp\left(\mu_{IC}^2 + \frac{1}{2}\Sigma_{11}^2 \right)}{\mu_E^1 - \mu_E^2} \tag{5.5}$$

In addition, the INB (λ) parameter is given by

$$\text{INB}(\lambda) = \lambda\left[\mu_E^1 - \mu_E^2 \right] - \left[\exp\left(\mu_{IC}^1 + \frac{1}{2}\Sigma_{11}^1 \right) - \exp\left(\mu_{IC}^2 + \frac{1}{2}\Sigma_{11}^2 \right) \right] \tag{5.6}$$

In the above expression, λ is the willingness-to-pay parameter. The ΔACER parameter has the expression

$$\Delta\text{ACER} = \frac{\exp\left(\mu_{IC}^1 + \frac{1}{2}\Sigma_{11}^1 \right)}{\mu_E^1} - \frac{\exp\left(\mu_{IC}^2 + \frac{1}{2}\Sigma_{11}^2 \right)}{\mu_E^2} \tag{5.7}$$

For the parameters given in Equations 5.5 through 5.7, the asymptotic mean and variance of the corresponding point estimates can be obtained using the delta method, and confidence limits can then be obtained using the asymptotic normality of the point estimates. Alternative approaches for interval estimation include application of the bootstrap (parametric or non-parametric), or Fieller's theorem based either on asymptotic normality or implemented using the bootstrap (the latter is referred to as the bootstrap-Fieller approach). We refer to Briggs et al. (1999), Fan and Zhou (2007), Mihay-lovaet al. (2011), Nixon et al. (2010), and Polsky et al. (1997) for a comparison of some of these methods.

5.2.4 The Generalized Pivotal Quantity (GPQ) Approach

In a recent paper, Bebu et al. (2016a) investigated the novel concept of a GPQ to obtain accurate confidence limits in the log-normal/normal scenario (or in the normal/normal scenario). In view of its novelty in the context of CEA, we shall briefly explain the GPQ approach for the interval estimation of the parameters in Equations 5.5 through 5.7. The GPQ approach exists due to Weerahandi (1993), and is a special case of the fiducial methodology in statistics. A GPQ for a parameter is defined in terms of random variables and their observed values, and a GPQ is required to satisfy two conditions: (1) when the observed data are fixed, the GPQ has a distribution that is free of any unknown parameters; and (2) when the random variables are replaced by their respective observed values, the GPQ is free of any nuisance parameters, and is often equal to the parameter of interest. In order to exhibit GPQs in the context of CEA, we consider Equation 5.5, and let $X_i^j = \left[ln\left(C_i^j\right), E_i^j \right]^T$ (T for transpose) denote a sample from $N\left(\mu^j, \Sigma^j\right)$, $i = 1, 2, \cdots, n_j$. Then, we write $\mu^j = \left(\mu_{lC}^j, \mu_E^j\right)$, $j = 1, 2$, and let $\bar{X}^j$ and $\hat{\Sigma}^j$, respectively, denote the sample mean vector and the sample covariance matrix based on the sample from the jth group, $j = 1, 2$, so that $\hat{\Sigma}^j$ is an unbiased estimator of Σ^j, $j = 1, 2$.

We shall also denote the observed values of $\bar{X}^j$ and $\hat{\Sigma}^j$ by $\bar{X}_o^j$ and $\hat{\Sigma}_o^j$, respectively. The GPQs for Σ^j and μ^j will be denoted by T_{Σ^j} and T_{μ^j}, respectively, and are given below. Here, we omit details of their derivations and refer the reader to Bebu and Mathew (2008) and Gamage et al. (2004) for further information. Consider

$$T_{\Sigma^j} = \left[W_2\left(\hat{\Sigma}_o^{j-1}, n_j - 1\right) \right]^{-1}$$

$$T_{\mu^j} = \bar{X}_o^j - \frac{1}{\sqrt{n}} \times T_{\Sigma^j}^{\frac{1}{2}} \times Z^j$$

(5.8)

$j = 1, 2$, where $Z^j \sim N\left(0, I_2\right)$, $W_2\left(\Sigma, m\right)$ denotes the bivariate Wishart distribution with $df=m$, and $T_{\Sigma^j}^{\frac{1}{2}}$ denotes the positive definite square root of T_{Σ^j}. A property of the GPQ that we shall use is that for any scalar function of μ^j and Σ^j, say $g\left(\mu^1, \mu^2, \Sigma^1, \Sigma^2\right)$, a GPQ is given by $g\left(T_{\mu^1}, T_{\mu^2}, T_{\Sigma^1}, T_{\Sigma^2}\right)$, and confidence limits for $g\left(\mu^1, \mu^2, \Sigma^1, \Sigma^2\right)$ can be obtained as the percentiles of $g\left(T_{\mu^1}, T_{\mu^2}, T_{\Sigma^1}, T_{\Sigma^2}\right)$.

The parameters ICER, INB (λ), and ΔACER given in Equations 5.5 through 5.7, respectively, are scalar valued functions of μ^j and Σ^j ($j = 1, 2$). Thus, the above property can be used to obtain confidence limits for these parameters. The required percentiles have to be obtained based on Monte Carlo simulation; however, the computations are quite straightforward. The computation can be carried out using the following algorithm, presented in the context of computing confidence limits for the ICER parameter: $\bar{X}^j$.

1. Compute $\bar{X}_o^j$ and $\hat{\Sigma}_o^j$, the observed values of $\bar{X}^j$ and $\hat{\Sigma}^j$, respectively, $j = 1, 2$, from the given sample.
2. Independently generate $\tilde{\Sigma}^j \sim W_2\left(\hat{\Sigma}_o^{j-1}, n_j - 1\right)$ and $Z^j \sim N(0, I_2)$, for $j = 1, 2$.
3. Now, compute the quantities $\Gamma^j = \tilde{\Sigma}^{j-1}$ and $\tilde{\mu}^j = \bar{X}_o^j - \dfrac{1}{\sqrt{n_j}} \Gamma^{j \frac{1}{2}} Z^j$.
4. Let $\tilde{\mu}^j = \left(\tilde{\mu}_1^j, \tilde{\mu}_2^j\right)$, and Γ_{11}^j denote the first diagonal element of Γ^j, $j = 1, 2$.
5. A GPQ for the ICER is now given by

$$T_{ICER} = \frac{\exp\left(\tilde{\mu}_C^1 + \dfrac{1}{2}\Gamma_{11}^1\right) - \exp\left(\tilde{\mu}_C^2 + \dfrac{1}{2}\Gamma_{11}^2\right)}{\tilde{\mu}_E^1 - \tilde{\mu}_E^2}$$

6. Repeat steps 2–5 k times, resulting in k values of T_{ICER}. A 95% confidence interval for ICER is given by the 2.5[th] and 97.5[th] percentiles of the GPQ T_{ICER}.

It should be clear that the algorithm can be easily modified for computing confidence limits for the parameters INB (λ) and ΔACER.

An attractive feature of the GPQ methodology is that it can address the interval estimation of an arbitrary function of the mean vectors and covariance matrices in the model (Equation 5.4). For example, suppose that, in addition to the costs, the effectiveness measures are also log-normally distributed. That is, we have the model

$$\begin{pmatrix} \ln(C^j) \\ \ln(E^j) \end{pmatrix} \sim N\left(\begin{pmatrix} \mu_{lC}^j \\ \mu_{lE}^j \end{pmatrix}, \Sigma^j \right)$$

where μ_{lE}^j is the population mean of the log-transformed effectiveness measures, $j = 1, 2$. Now, the ICER and INB (λ) have the expressions

$$ICER = \frac{\exp\left(\mu_{lC}^1 + \dfrac{1}{2}\Sigma_{11}^1\right) - \exp\left(\mu_{lC}^2 + \dfrac{1}{2}\Sigma_{11}^2\right)}{\exp\left(\mu_{lE}^1 + \dfrac{1}{2}\Sigma_{22}^1\right) - \exp\left(\mu_{lE}^2 + \dfrac{1}{2}\Sigma_{22}^2\right)},$$

$$INB(\lambda) = \left[\exp\left(\mu_{IC}^1 + \frac{1}{2}\Sigma_{22}^1 \right) - \exp\left(\mu_{IC}^2 + \frac{1}{2}\Sigma_{22}^2 \right) \right]$$

$$- \left[\exp\left(\mu_{IC}^1 + \frac{1}{2}\Sigma_{11}^1 \right) - \exp\left(\mu_{IC}^2 + \frac{1}{2}\Sigma_{11}^2 \right) \right],$$

where Σ_{22}^j is the second diagonal element of the covariance matrix Σ^j, $j = 1$, 2. The ΔACER parameter can also be similarly modified. It should be clear that the GPQ methodology can be easily adopted in this scenario in which the costs and effectiveness are both lognormally distributed.

In their recent work, Bebu et al. (2016a) have reported fairly extensive numerical results with respect to the coverage probability of the GPQ-based confidence limit, along with those based on the bootstrap-Fieller (B-Fieller) approach, the delta method, and the parametric and nonparametric bootstrap approaches, under the lognormal/normal model (Equation 5.4). In terms of maintaining the coverage probability, the GPQ method is noted to outperform all of the other approaches. An important aspect to note is that the GPQ-based confidence interval for the ICER will always be a finite interval, while it is well known that the Fieller approach does not always produce a finite interval. The following example, discussed in Bebu et al. (2016a), further illustrates this aspect.

5.2.5 An Example

The example is taken from Willan and Briggs (2006), and the data were from a trial of prostate cancer therapy. The same data set was also analyzed in Bebu et al. (2016a). In the trial, 114 symptomatic, hormone-resistant prostate cancer subjects were randomized to two groups, specifically a group that received prednisone alone, and a second group that received prednisone and mitoxantrone. The respective sample sizes were $n_1 = 61$ and $n_2 = 53$. The subjects were followed until death. The costs were in Canadian dollars (C\$), and the effectiveness measure used was quality-adjusted life-weeks. Some summary statistics were given in Willan and Briggs (2006), and the sample means gave average costs of C\$29,039 and C\$27,039, along with average effectiveness of 28.11 and 40.89, for prednisone and mitoxantrone, respectively.

A willingness-to-pay amount of C\$500 was used. The data indicated that the prednisone and mitoxantrone were less costly and more effective as compared with prednisone alone. The purpose of our data analysis is to illustrate our methodology and to point out possible differences among the different methods for computing confidence limits for ICER and INB (λ). In this context, the results are intended to be used for illustrative purposes, and not to draw any conclusions or comparisons about the safety or efficacy of the treatment regimens. It was noted in Bebu et al. (2016a) that, in this example, both the costs and effectiveness measures were lognormally distributed. Thus we shall use the corresponding formulas for the ICER and INB (λ). The following tables give the 95% and 90% confidence limits for ICER and INB (λ), as they correspond to $\lambda = 500$. In the table, "B-Fieller" corresponds to the interval obtained using Fieller's theorem, except that the required percentiles were obtained by a parametric bootstrap method. Furthermore, "N-Boot" corresponds to the interval obtained using the nonparametric bootstrap (Tables 5.1 and 5.2).

We note that for the ICER, the 90% confidence intervals based on the three methods agree somewhat. Considering the 95% intervals, we note that the B-Fieller interval is a union of two disjoint segments. As already noted, the GPQ will always provide a finite interval. We also note significant differences between the two intervals for INB (λ). The numerical results in Bebu et al. (2016a) show that the coverage probabilities of the N-Boot confidence interval for INB (λ) are very often unsatisfactory; they can be significantly lower than the nominal level. However, the GPQ-based intervals were always satisfactory. Our overall conclusion is that the GPQ-based confidence intervals can be recommended for practical use. Their coverage probability performance is very satisfactory, and they always produce a finite interval for the ICER parameter.

In their work, Bebu et al. (2016a) have also provided an R code for performing the computations necessary to obtain the GPQ-based confidence limits. While we acknowledge that the GPQ-based methodology may be

TABLE 5.1

Confidence Intervals for the Incremental Cost-Effectiveness Ratio

	GPQ	B-Fieller	N-Boot
95% interval	(−2018, 1070)	(−3643, 852)$\cup$(2962, ∞)	(−1971, 807)
90% interval	(−1218, 424)	(−1266, 453)	(−1216, 438)

Source: Reprinted from *Comput Stat Data Anal.*, Vol. 94, Bebu, I. et al., Parametric cost-effectiveness inference with skewed data, 210–220, Copyright (2016a), with permission **Elsevier B.V.**

TABLE 5.2

Confidence Intervals for Incremental Net Benefit ($\lambda = 500$)

	GPQ	N-Boot
95% interval	(42, 26936)	(−350, 16606)
90% interval	(1903, 23898)	(1041, 15225)

Source: Reprinted from *Comput Stat Data Anal.*, Vol. 94, Bebu, I. et al., Parametric cost-effectiveness inference with skewed data, 210–220, Copyright (2016a), with permission **Elsevier B.V.**

new to researchers in CEA, the availability of its R code, and the accuracy of its coverage probability performance, makes it a viable approach worth considering.

5.2.6 A Probabilistic Measure of Cost-Effectiveness

Recently, Bebu et al. (2016b) introduced some probabilistic measures of cost-effectiveness that may be used to conceptualize the relative performance of an intervention at the individual level. The probabilistic measures can be used to assess how likely the first treatment is to be less costly and more effective in comparison with the second one, which we will term to be "cost-effective" for this exposition. It is equivalent to being "dominant," a term often used in the cost-effectiveness literature, as well as to being (in the more general sense) cost-effective when the willingness to pay for a health gain is zero (although this can be generalized somewhat, as explained below).

The cost-effectiveness measure proposed by the authors is termed the cost-effectiveness probability (CEP), and is the probability for the first treatment to be less costly and more effective as compared with the second treatment. An attractive feature of the proposed CEP parameter is that it is invariant with respect to the monotone transformations of Cost and Effectiveness, a property not shared by the parameters ICER, INB, ACER and CEAC. Consequently, as noted in Manning (1998), the interpretation of the parameters ICER, INB, ACER and CEAC on the original scale, rather than on the transformed scale, is difficult.

In order to define the CEP parameter, we note that for two randomly selected subjects, one from each treatment group, the first treatment is more cost-effective than the second treatment if

$$C^1 \leq C^2 \ \& \ E^1 \geq E^2 \tag{5.9}$$

It should be clear that Equation 5.9 corresponds to the southeast quadrant of the cost-effectiveness plane. The cost-effectiveness probability (CEP) is simply the probability for the event defined in Equation 5.9:

$$CEP = P\left(C^1 \leq C^2, E^1 \geq E^2\right) \tag{5.10}$$

As already noted, a desirable feature of Equation 5.10 is that it is invariant with respect to the monotone transformations of the cost and effectiveness. Therefore, when the cost and effectiveness are both continuous, one can assume that the random variable (Cost, Effectiveness) follows a bivariate normal distribution, under the assumption that transformation to normality is possible. From the definition given in Equation 5.10, it should be clear that large values of the CEP indicate the cost-effectiveness of the first treatment as compared with the second.

Another attractive feature of the CEP is that it can be easily modified to reflect various practical scenarios involving cost-effectiveness comparisons. As an example, suppose the first treatment is expected to be more effective, but also more costly up to a margin, say δ_C (similar to a willingness-to-pay parameter). The CEP can be modified to reflect this scenario as follows:

$$\text{CEP}(\delta_C) = P\left(C^1 \leq C^2 + \delta_C, E^1 \geq E^2\right)$$

Now, large values of CEP (δ_C) indicate the cost-effectiveness of the first treatment as compared with that of the second. One can also include a second margin δ_E for the effectiveness, so as to capture increased effectiveness of a certain magnitude:

$$\text{CEP}(\delta_C, \delta_E) = P\left(C^1 \leq C^2 + \delta_C, E^1 \geq E^2 + \delta_E\right)$$

In practical terms, δ_E can be seen as the minimum clinically meaningful difference, while δ_c can be seen as the maximum tolerable budget impact. This formulation allows for nonzero willingness to pay for a health gain, albeit in a noncontinuous sense. Other possibilities suggested in Bebu et al. (2016b) include conditional probabilities, such as $P(C^1 \leq C^2 \mid E^1 \geq E^2)$. Such a conditional probability is relevant for investigating the extent to which the first treatment is less costly among the subjects for whom it is more effective. A modified parameter suggested by the authors, denoted by ΔCEP, is defined as

$$\Delta\text{CEP} = P\left(C^1 \leq C^2, E^1 \geq E^2\right) - P\left(C^1 \geq C^2, E^1 \leq E^2\right)$$

The motivation for defining ΔCEP is that, apart from the probability in (Equation 5.10), the probability $P(C^1 \geq C^2, E^1 \ E^2)$ can also be significantly different from zero. Clearly, the latter probability represents the proportion of patients for whom the first treatment is more costly and less effective as compared with the second. In view of this, a positive value of ΔCEP indicates that treatment 1 can be considered cost-effective as compared with treatment 2; a negative value indicates the converse. Thus, it is not difficult to show that $\Delta\text{CEP} = P\left(C^1 \leq C^2\right) - P\left(E^1 \leq E^2\right)$.

Assuming that transformation to normality is possible, let us consider the case when (Cost, Effectiveness) follows a bivariate normal distribution:

$$\begin{pmatrix} C^j \\ E^j \end{pmatrix} \sim N\left(\mu^j, \Sigma^j\right) \tag{5.11}$$

$j = 1$, 2. Write $\mu^j = \left(\mu_C^j, \mu_E^j\right)'$, $\Sigma^j = \Sigma_{ll'}^j$, for i, l, $l' = 1$, 2. The CEP parameter can now be simplified as

$$\begin{aligned}
\text{CEP} &= P\left(C^1 - C^2 \leq 0 \ \& \ E^1 - E^2 \geq 0\right) \\
&= P\left(C^1 - C^2 \leq 0\right) - P\left(C^1 - C^2 \leq 0 \ \& \ E^1 - E^2 \leq 0\right) \\
&= \Phi\left(0; \mu_C^1 - \mu_C^2, \Sigma_{11}^1 + \Sigma_{11}^2\right) - \Phi_2\left((0, 0)' ; \left(\mu_C^1 - \mu_C^2, \mu_E^1 - \mu_E^2\right)', \Sigma^1 + \Sigma^2\right)
\end{aligned} \tag{5.12}$$

In the above expression, Φ and Φ_2 are used to denote, respectively, the cumulate distribution functions of the univariate normal and bivariate normal distributions. One can similarly obtain a representation for ΔCEP. Such representations facilitate the point estimation of CEP and ΔCEP based on a random sample from the bivariate normal distribution in Equation 5.11. Furthermore, a large sample inference can be obtained using the delta method. Since CEP and ΔCEP are both functions of the bivariate normal parameters, the GPQ methodology can also be employed for obtaining confidence limits. Details are provided in Bebu et al. (2016b). Based on numerical results, the authors conclude that the GPQ approach provides accurate coverage probabilities, regardless of the sample size. Apart from the bivariate normal scenario, the authors have also considered inference concerning the probabilistic measures, based on U-statistics.

We now return briefly to the example discussed earlier and reported in Bebu et al. (2016b). Based on the data, it was already noted that the use of prednisone and mitoxantrone was estimated to be less costly and more effective as compared to prednisone alone. For this example, we estimated the CEP using the expression given in Equation 5.12, and also computed a 95% confidence interval using the GPQ methodology. The point estimate turned out to be 0.2366, and the 95% confidence interval is the interval (0.1693, 0.3099). In other words, we conclude that, as compared with prednisone alone, prednisone plus mitoxantrone was estimated to be a more effective and less costly treatment option among 23.66% of the population, with a corresponding interval estimate given by (0.1693, 0.3099). We note that the CEP does provide information that is not available from the traditional measures ICER and INB (λ). We refer to Bebu et al. (2016b) for a more detailed analysis of this example.

Clearly, the well-established traditional approaches will continue to play a leading role in the evaluation of cost-effectiveness. The GPQ methodology due to Bebu et al. (2016a,b) provides accurate inference for the traditional parameters such as ICER and INB. The CEP parameter is not meant to replace the traditional parameters. As should be clear from the example, the CEP is meant to supplement the traditional parameters in view of the fact that no single measure can adequately characterize the cost-effectiveness comparison between two treatments.

Here, we have not considered the scenario in which covariates could be present and affecting cost and/or effectiveness. In addition, we have not addressed CEA based on observational data. When covariates are present, models that incorporate them are necessary while analyzing the data from randomized clinical trials or observational studies; in the latter case, it will adjust for confounders. In the work of Bebu et al. (2016a), models involving covariates are indeed considered. Furthermore, covariate-adjusted estimators are given in Willan and Briggs (2006). In addition, based on covariates, it may be possible to identify patient subgroups for whom a health care intervention is most or least cost-effective.

Cost-effectiveness analysis based on cluster randomized trials (CRTs) is also a problem of considerable practical interest. These are trials in which the unit of randomization is a cluster (rather than individual patients), for example, a hospital, a primary care physician, or an insurance carrier. Models and approaches for CEA in the context of CRTs, and case study results in the context of such trials, have been discussed in the literature; see, for example, the work of Gomes et al. (2012a,b). In Gomes et al. (2012a), seemingly unrelated regression models and multi-level models that include random effects are proposed, and the authors have also used the bootstrap method in order to obtain confidence limits for the INB(λ).

In Gomes et al. (2012b), the authors note that "[CEAs] that use data from [CRTs] can face particular methodological challenges. These studies require methods that address clustering in both costs and outcomes, recognize the correlation between individual- and cluster-level costs and outcomes, and make appropriate assumptions about the distribution of these endpoints."

In the same article, the authors conducted a conceptual review, including studies for CEA in CRTs published from 1997 to 2009, and noted that the methods used in the vast majority of these studies fail to address the statistical challenges. It should be noted that multi-level models that include random effects, similar to those considered in Gomes et al. (2012a), are also relevant while investigating the transferability of economic data across jurisdictions, since variability across locations is to be expected in the cost-effectiveness of health interventions. Also see the International Society for Pharmacoeconomics and Outcomes Research (ISPOR) Good Research Practices Task Force Report by Drummond et al. (2009), the article by Briggs (2010), and the ISPOR Research Task Force Report by McGhan et al. (2009). We shall not go into further technical details, except to note that these are topics that require further investigation.

5.3 Incorporating Data from Indirect Comparisons and Observational Studies

As pointed out earlier, a major issue in CEA concerns the availability of reliable data from RCTs, which are accepted as the gold standard for evidence-based medicine. While CEA can be conducted alongside individual RCTs

(Ramsey et al., 2015), in many situations, RCTs may not employ the comparators or populations of interest. In other cases, the data from RCTs may not be adequate, since the conditions under which such trials are conducted may not be reflective of the real-world situations in which the drugs are to be used. Accordingly, it may be essential to resort to indirect ways of obtaining the desired information.

When head-to-head comparative data from RCTs are not available for treatments that need to be studied, in comparative effectiveness research, the issue is often handled using indirect comparison procedures (Bucher et al., 1997; Lu and Ades, 2004). Another potential approach to fill the data gap is to use pertinent effectiveness data from observational studies (Berger et al., 2009). Also, given individual patient data from one comparative RCT, and summary data from the literature for related treatment comparisons, a matching-adjusted approach for indirect comparisons has been proposed (Signorovitch et al., 2010). However, despite the appeal of these procedures to serve as viable options in the absence of direct comparative evidence from RCTs, they also have certain inherent shortcomings that require careful scrutiny. In the following, we outline the issues associated with some of the commonly used techniques and measures that need to be taken, with particular emphasis on network meta-analysis (Lumley, 2004) and the use of data from observational studies (Cox et al., 2009) in relation to CEA. It should be noted that some of the issues raised are complex and may not have definitive solutions. However, given the potential implications for decision-making relating to health care resource utilization, it is important that adequate attention be paid to these fundamental issues.

5.3.1 Indirect Comparisons

Consider the problem of performing CEA involving treatment groups A and C, which have never been studied in a head-to-head RCT. Suppose, however, that data from similar RCTs are available with respect to comparing each treatment to a common control B, and give the estimated effects, $d_{AB} = E_A - E_B$, and $d_{CB} = E_C - E_B$, respectively. Bucher et al. (1997) proposed estimating the effect of A relative to C indirectly by

$$d_{AC} = d_{AB} - d_{CB}$$

For inference, the standard error (SE) of the resulting estimator may be computed as

$$SE(d_{AC}) = \sqrt{SE(d_{AB})^2 + SE(d_{CB})^2}$$

The allure of the approach lies in the fact that it preserves some of the benefits of randomization, as compared with the naïve approach based on the direct treatment effect estimates E_A and E_C. In typical applications, it may also be desirable to synthesize information from both direct sources and indirect sources. In these situations, one may employ network meta-analysis or mixed treatment comparison techniques (Caldwell et al., 2005).

Consider, for example, a situation involving four treatment groups *A*, *B*, *C* and *D*, studied in a *connected* network of RCTs—that is, every treatment has been studied with at least one of the other treatments in the same RCT. Following the convention of Hoaglin et al. (2011), we assume that treatment *A* is used as a reference group, and let η_{jk} denote the outcome for the jth treatment in the kth trial. Designating one treatment, *b*, as the base treatment in each study, the network is formulated such that the base treatments come "after" *A*, and the nonbase treatments come "after" all the base treatments in the alphabet. The underlying fixed effects model may then be specified as follows:

$$\eta_{jk} = \begin{cases} \mu_{jb}, & b = A, B, C, \text{ if } k = b, \\ \mu_{jb} + \delta_{bk}, & k = B, C, D \text{ if } k \text{ is "after" } b \end{cases}$$

where μ_{jb} is the mean treatment effect for treatment *j* in trial *b*, and

$$\delta_{bk} = \delta_{Ak} - \delta_{Ab}$$

is the fixed effect of treatment *k* relative to treatment *b*, *with* $\delta_{AA} = 0$. In most applications, the analysis is executed using mixed effects models in a Bayesian framework, where the treatment effects are given vague priors. For further details, see Chapter 6 of this monograph or Hoaglin et al. (2011). Because of the convenience provided by the Bayesian approach to present probabilities of relative effect sizes, this method appears to be more popular among practitioners.

There are several package programs to implement network meta-analysis or MTC. Most of the available software programs require customizing preexisting codes. Many analysts tend to prefer R, because of its desirable features, including graphic capabilities and interfacing with other common software programs such as WinBUGS (MRC and Imperial College of Science, Technology and Medicine). A review of available packages may be found in Neupane et al. (2014).

Incidentally, these approaches have major limitations, including those that are commonly encountered in standard meta-analysis, such as heterogeneity and confounding, as well as others that are inherent in the underlying construction of the procedures. One critical issue pertains to the assumption that the trials contributing evidence are *similar* with respect to relevant attributes, including patient characteristics, study design, and health care practices.

In addition, when performing mixed treatment comparisons, it is also essential to assume that there is *consistency* in the evidence generated from the direct and indirect sources. It should be noted that the assumption of similarity is generally not testable, since its underlying requirement is that the same effect size would have been obtained had each trial compared the treatments studied in the other study. Although direct testing measures are not available, there are indirect qualitative and quantitative approaches that may be used to add a level of credibility to the results (see, e.g., Alemayehu, 2011). Further, when results from direct and indirect sources are inconsistent, it is important to investigate the possible causes of discrepancy, and to take appropriate measures. For discussions of these issues and of mesures that need to be taken, see Bucher et al. (1997), Dias et al. (2010), and Dias et al. (2013).

There has been sizable growth in the body of literature pertaining to the use of indirect evidence in CEA. For example, Khoo et al. (2015) used network meta-analysis to compute the relative efficacies of several antidepressants and to apply the efficacy results in CEA based on decision trees. Other examples include Coyle et al. (2014) and Zhao et al. (2016).

5.3.2 Effectiveness Data from Observational Studies

In the absence of adequate effectiveness data from RCTs, observational studies may be used to fill the evidentiary gaps (Berger et al., 2009). While such data are known to have relatively high external validity, there are major limitations that need to be recognized and addressed. Specifically, the use of data from non-RCT sources may lead to biased results as a consequence of confounding, absence of blinding, and lack of consistent requirements for patient follow-up or prespecifications of procedures and outcome measures. It is therefore critical to investigate the potential sources of bias, and to take appropriate measures to minimize the impact on CEA.

For measured confounders, there are commonly used techniques to control for their effects, including the usual analysis of covariance or other techniques, such as propensity score analysis (Rosenbaum and Rubin, 1983). Although approaches have been proposed to handle unmeasured confounders, such as instrumental variables, there is no universally accepted procedure that has been shown to give reliable results. Further, there is also the problem of confounding by indication, which is common in drug safety studies. This type of confounding arises when the indication is also a risk factor for the outcome. In this case, there is always the risk of residual confounding (Salas et al., 1999), even when all attempts are made to control for observed confounders.

In recent studies, observational study results have been shown to be dependent on the choice of database as well as on analytical and design strategies. In one study, for example, Madigan et al. (2013a) noted that, depending on the observational database selected, analytical results may vary from one extreme to another. In a related study, Madigan et al. (2013b) also showed that the choice of study design and analytical techniques could also dramatically influence the study conclusions, if all other factors remained constant.

Thus, while observational data may play a critical role in CEA, as a source of comparative evidence complementary to RCTs, caution should be exercised in the use of such data in the calculation of ICERs and other similar metrics. In addition to the measures summarized above in the context of controlling for bias, best practices should be adopted, particularly those intended to strengthen the values of data from secondary sources (Cox et al., 2009; Johnson et al., 2009). See also Faria et al. (2015) for suggested guidelines on the use of data from non-RCT sources in CEA.

5.3.3 Issues with Cost Data

Cost data are inherently observational and are typically associated with issues that require special handling in modeling and analysis. For example, the usual assumptions for ordinary least squares (OLS) may not be satisfied, as cost data may be heteroscedastic, skewed, or inflated with zeros. Uncritical use of the usual OLS approaches in analyzing cost data is, therefore, likely to offer biased results and, hence, unreliable and misleading CEA findings.

A common approach to dealing with heteroscedasticity and nonnormality is the use of suitable transformations, including Box–Cox, with log transformation being a popular choice. However, with simple transformations of a cost variable Y_i, the predicted cost may be affected by retransformation bias, often requiring correction using appropriate techniques. For example, when $Z_i = log(Y_i)$ is assumed to be normal with mean μ_z and variance σ^2, it is known that

$$E[Y] = \exp\left(\mu_z + \frac{\sigma 2}{2}\right)$$

Clearly, the log scale introduces bias if not properly retransformed. One approach is to use Duan's (1983) smearing estimator.

In the regression context, suppose $Z_i = g(Y_i) = x_i\beta + \varepsilon_i$, where the ε_i are identically and independently distributed error terms, with $E(\varepsilon_i) = 0$ and $Var(\varepsilon_i) = \sigma^2$, so that

$$Y_i = g^{-1}(x_i\beta + \varepsilon_i) = h(x_i\beta + \varepsilon_i)$$

β is estimated by $\hat{\beta} = (X'X)^{-1}X'Z$. Obviously, $E(Y \mid x) = E[h(x'\beta + \varepsilon)] \neq E[h(x'\beta)]$. For arbitrary F,

$$E(Y \mid x) = E[h(x'\beta) + \varepsilon)] = \int h(x'\beta)dF(\varepsilon)$$

Estimating F by the empirical cumulative distribution function of the estimated residuals, we get $F_n(e) = n^{-1}\Sigma_i I[_i \leq e]$, where $\hat{\varepsilon}_i = z_i - x_i\hat{\beta}$, and $I[\cdot]$ denotes the indicator function. Then, the mean of the untransformed data may be estimated by

$$\hat{E}(Y \mid x) = \int h\left(x'\hat{\beta}\right)F_n\left(\hat{\varepsilon}_i\right)$$

$$= n^{-1}\sum_{i=1}^{n} h\left(x'\hat{\beta} + \hat{\varepsilon}_i\right)$$

For log transformation, the estimated mean may be computed as

$$\hat{E}(Y \mid x) = n^{-1} \sum_{i=1}^{n} \exp\left(x'\hat{\beta} + \hat{\varepsilon}_i\right)$$

$$= \exp\left(x'\hat{\beta}\right) n^{-1} \sum_{i=1}^{n} \exp\left(z_i - x_i'\,\hat{\beta}\right)$$

However, Duan's approach is often difficult to accomplish in heteroscedastic cases, especially when the number of regressors is large and contains continuous measures. Therefore, it may be appropriate to use models that incorporate the variance structure and that do not heavily rely on the usual linear model assumptions. When the relationship between the variance and the mean of cost, conditional on the covariates, can be specified, a generalized linear model (GLM) with an appropriate link function may be a viable option (Manning and Mullahy, 2001).

A more common problem in the analysis of cost data is the proper handling of the situation involving many zero values. This occurs, for instance, when some of the participants in a health plan did not incur any cost. In this case, one popular approach is to model $ln(Y+c)$ for a suitably chosen positive constant c. One issue with this approach, however, concerns the fact that zero values may be associated differently with the covariates of interest. In addition, there is no clear way of choosing c, and retransformation may be quite problematic (Duan et al., 1983). Other approaches for handling the problem, also used for skewed data, include gamma and Box–Cox regression models (Manning et al., 2005). When there are too many zeros, gamma regression tends to be unstable. In the case of Box–Cox regression, the model parameters may not be readily interpretable or not always easy to identify.

One approach that is popular among practitioners is the so-called two-part model, in which Part 1 of the model involves a binary outcome—that is, whether or not a participant incurred any cost—while Part 2 deals with the regression of the nonzero cost outcomes on the covariates. More specifically, supposing that Y is the outcome variable and X a vector of predictors, then the expected value of Y conditional on X may be given by

$$E(Y \mid X) = P(Y > 0 \mid X)E(Y \mid Y > 0, X)$$

The binary part of the model is handled using a logistic regression or probit model, which gives the predicted probability that the dependent variable is nonzero given the covariates. In Part 2, observations with nonzero cost are regressed on the predictors using a suitable GLM or OLS (with log cost) model.

Obviously, the above approach results in separate parameter estimates for the two components—one for participation and the other for the conditional quantity—each parameter having its own policy relevance. With this approach, there is no easy analytical way of constructing confidence intervals;

consequently, bootstrap or other resampling techniques are commonly used instead. For a discussion of the two-part model in health economic analysis, see Mullahy (1998).

An alternative approach that permits modeling heterogeneity, depending on patient characteristics or outcome values, is based on the so-called finite mixture models (see, e.g., Deb and Trivedi, 2002). The approach involves pre-specifying the number of classes (m) and mixing probabilities ($\pi_j : j=1, \ldots, m$). More specifically, suppose the probability density function f_j corresponding to the jth class can be specified. Then,

$$E[Y \mid X] = \sum_{j=1}^{m} \pi_j E(Y \mid X, f_j)$$

Estimation is usually executed using the maximum likelihood procedure. In practice, it is often challenging to specify m, the number of components, except in situations in which the definition of components is straightforward (e.g., male versus female) or when there is adequate historical data to inform the choice. In addition, the likelihood function may have multiple local maxima, leading to numerical and computational difficulties.

5.4 Decision Analysis

Decision analysis is a broad field with roots and applications in many areas, from business to military strategy. A major step in its formalization was the work of von Neumann and Morgenstern (1953) in game theory and the formulation of expected utility. The expression of the expected utility, or benefit, of a strategy—such as a medical treatment—as the sum, over all possible outcomes of treatment, as the "utility" of an outcome (see Section 5.4.2) times the probability of the occurrence of that outcome, provides a straightforward basis, along with similarly constructed expected costs, for evaluating that strategy. Typically, decision analysis in health care involves the construction of a mathematical model, possibly simple and possibly quite complex, to evaluate the cost-effectiveness of a treatment based on certain probabilities, utilities, and costs (each one either known, estimated, or assumed), rather than by direct observation of individuals in a trial or undergoing a treatment in the real world.

In certain applications, decision analysis may be appropriate when treatment options need to be compared in patients that change health states over time. Use of models in such cases is particularly important when there is inadequate information from clinical trials or other sources to address a specific situation. For example, data from clinical trials may have limited value, since trial duration may be short, certain subgroups may have been excluded from consideration, or the trial may have been conducted in a restrictive clinical setting. Thus, models may serve a useful purpose to generalize trial results to wider population groups or to different clinical settings, and to extrapolate results to time points beyond the period mandated by the study protocol. In addition, these models are important vehicles for adding costs to health care events and health states, enabling health economic analyses of treatment.

In decision analysis, there are several commonly used approaches. Examples include decision trees (Revicki et al., 1995), Markov models (Briggs and Sculpher, 1998), or discrete event simulation (Karnon and Brown, 1998). Decision trees are widely used for simple cases that involve relatively short time periods and, as such, have the advantage of simplicity and relatively less reliance on heavy assumptions. Nevertheless, for more complex situations involving progressive and recursive disease profiles, Markov models are often the preferred techniques in health economic analysis. Although not widely used to date in health economics, due largely to its lack of transparency, discrete event simulation may also be more suited in situations requiring flexibility of the modeling approach. In what follows, we consider some of the theoretical and practical aspects of commonly used models, with an emphasis on decision trees and Markov models, and shed light on their limitations and underlying assumptions.

5.4.1 Steps in Decision Analysis

A critical step in decision analysis is a clear definition of the ultimate goal and intended conclusion. The goal may involve maximizing treatment outcome or minimizing cost, or it may be a function of both. Assuming that there is a finite number of clearly specified decision alternatives, one then needs to list all of the possible health outcomes for each decision alternative, as well as the relevant timeframe for and the probability of each outcome. If interest lies in the calculation of ICERs, one needs to compute the expected health outcomes and costs separately at the time of the treatment decision. Since the computation is cumbersome, involving summing the products of health outcomes or costs in each stage, multiplied by the corresponding probabilities, the implementation of the approaches may require the use of customized software. A survey of alternative software packages may be found in Tosh and Wailoo (2008).

5.4.2 Outcome Measures

There are alternative outcome measures that are used in decision analysis. Examples include survival probability, quality-adjusted life-years (QALYs, a person's life expectancy as weighted by health state utilities), and cost of health resource utilization. Utility is most often measured by a multi-attribute utility scale (MAUS), often having a range of 0 to 1 (or 100), with 0 typically denoting "death," and 1 (or 100), perfect health (Hunink et al., 2001), though negative values are sometimes used for states rated "worse than death." Other alternative techniques for quantifying utility include the visual analog scale, standard gamble (Gafni, 1994), and time tradeoff (Burstrom et al., 2006) approaches.

In the *standard gamble* approach, study subjects are made to choose between their current health state versus a gamble with a binary outcome of perfect health (with probability π) and immediate death (with probability 1-π). A health state in which the participant is indifferent to the choices offered (i.e., the health state versus the gamble) is assigned a utility value of π. By contrast, the *time tradeoff* approach is based on a choice between life in a given health state (say t_1 years) relative to a shorter life in perfect health (t_2 years). The desired utility is given by ratio t_2/t_1.

The most commonly used MAUS, EQ-5D, is a five-item questionnaire with multiple choice responses for the severity of various aspects of health. The utility of the resulting health state combinations is then assigned based on published utility ratings from separate studies, often employing the standard gamble or time tradeoff (Rabin and de Charro, 2001).

The standard gamble, based on the fundamental axioms of utility theory (von Neumann and Morgenstern, 1953) is considered the most theoretically correct approach, and the time-tradeoff approach has also been found to be relatively valid (Torrance, 1976), though both are vulnerable to certain biases during the evaluation process (Dolan et al., 1996). Both the standard gamble and time-tradeoff are difficult to administer in large trial or population settings, so a MAUS is used much more often in practice.

To compute a QALY value, one can multiply the duration of a given health state (e.g., the number of years of life expectancy in the health state) by the corresponding utility value. A total value for an individual is obtained by summing the weighted utilities over the series of potential health states for the relevant time period.

On the other hand, costs for different states are determined as functions of several variables, including direct and indirect costs incurred as a consequence of a specific decision option. Direct costs may include costs of medicines, fees for doctors' visits, procedures, and hospital costs, while indirect costs may involve such costs as lost wages or nonmedical expenses incurred as a result of the intervention.

For summing costs and QALYs for time periods over a year, we need to use a *discounting factor* to account for the time value of money (e.g., the foregone interest) and the general preference for benefits occurring in the present rather than the future. The incorporation of indirect costs and discounting ensures that relative cost-benefits of alternative treatment options are computed with due regard to societal perspectives (Weinstein et al., 1996).

5.4.3 Decision Trees

Decision trees consist of nodes and branches that correspond to alternative decision options, as well as pathways of health outcomes and/or treatment options, and permit the combination of information from various sources. A simple decision tree is illustrated in Figure 5.1, consisting of *decision* and *chance nodes,* represented by a square and a circle, respectively. For each chance node, it is assumed that all possible outcomes could occur, and hence the corresponding probabilities must add up to 1. These probabilities play an important role in simulations to determine the proportion of patients at each node. Since time is not explicitly modeled in this approach, it is generally expected that the outcome measure (e.g., QALY) should account for the amount of time that elapsed between events. In most applications, it is thus customary to associate outcome measures, such as QALYs, with the endpoints of a tree, rather than the chance nodes. On the other hand, costs may vary within the various components of a tree—that is, some costs are associated with specific events, while others like QALYs may be modeled as accumulating over time—and hence, the total cost calculation involves summing up the costs associated with the different pathways, as weighted by the probability of each pathway.

In the illustrative example, the expected values for each decision option (i.e., treatment 1 versus treatment 2) are obtained through a process sometimes dubbed "folding back" the tree (Weinstein and Fineberg, 1980). The utility at each endpoint of the tree is multiplied by the respective probabilities and then added.

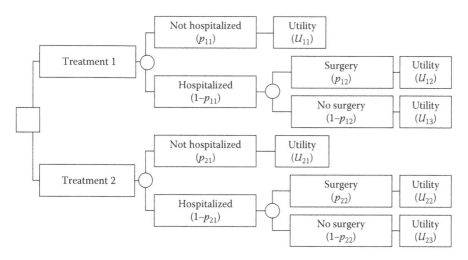

FIGURE 5.1
Illustration of a simple decision tree.

5.4.4 Markov Models

While decision analytic models are useful for simple or temporary health conditions, fully representing chronic conditions that require many periods of treatment can be cumbersome with such models. Markov modeling allows for a more systematic and efficient way to capture repeated cycles of treatment that have the potential to move patients across health states with their associated utilities and costs.

To fix ideas, consider a particular disease with s health states (e.g., asymptomatic, severe, death) (Figure 5.2). The definition of the states is a key step, and should be based on clinical and economic considerations. It is assumed that the states are mutually exclusive, and, in addition, it is customary to include an absorbing state, which is a state from which a patient cannot move out (e.g., death).

The movement of a patient's health from one state to the next is associated with transition probabilities. The probabilities may be obtained from alternative sources, including clinical trials, real-world data, literature review, mathematical models, and/or expert opinions. In Markov models, one is interested in the state of a patient at a given point in time, usually discretized into periods called "cycles." Depending on the disease and treatment modality, the length of the cycles may be days, months, years, or any fixed interval of time.

In the special case of Markov chain models, transition probabilities are assumed to be constant over time. This assumption permits representing the probabilities in each cycle using an $s \times s$ matrix P, given by

$$P = \begin{matrix} p_{11} & p_{12} & \cdots & p_{1s} \\ \vdots & \vdots & & \vdots \\ p_{s1} & p_{s2} & \cdots & p_{ss} \end{matrix}$$

where p_{ij} denotes the probability of transitioning from state i to state j. In most applications, it is also customary to impose the additional restriction that the probability at the kth cycle depends only on the health state at the $(k-1)^{th}$ cycle—also referred to as the memoryless property. However, the memoryless property may not always be a realistic assumption. For example,

FIGURE 5.2
A schematic diagram of health states and transition probabilities.

in some disease situations a fixed sequence of states called "tunnel states" may be needed; this depends on the incorporation of experiences of previous cycles (see, e.g., Hawkins et al., 2005). More general incorporation of memory is allowed in "semi-Markov" microsimulation models. These models have the same general cycle and health state structure as Markov models, but can calculate transition probabilities, utilities, and costs as a function of individual values—generally based on a formula, such as a risk score—that are updated as the person progresses through the simulation (Foucher et al., 2006).

Occasionally, information may be available with regards to the transfer rate (r) over a period of time t. In this case, under the assumption that the transition probability over the next time period does not depend on how long the patient has been in the present state, the transition probability may be computed as (Beck and Pauker, 1983)

$$p = 1 - e^{-rt}$$

In application, two types of Markov simulation models are used to assess health care programs, namely *cohort* simulation and *Monte Carlo* (or *individual*) simulation.

In a Markov cohort simulation, a cohort of patients is followed over a series of cycles. Beginning with a cohort of patients, initially distributed in different states, the model tracks the distributions of patients in subsequent cycles as they transit from one state to another. In this approach, transitions are often assumed to occur at the end of each cycle.

Alternatively, the transitions may be assumed to be uniformly distributed over the cycle interval, and a half-cycle correction may then be applied. This is often advised when the number of cycles is small and the length of cycles is too long and cannot be shortened (Barendregt, 2009).

To calculate health outcomes and costs, one needs to compute the respective values for each cycle and then aggregate over all of the cycles. More specifically, suppose there are c cycles, and s health states. For the ith cycle, assume there are n_{ij} patients in the jth state, with corresponding health outcomes u_{ij}. The total health outcome in cycle i is given by

$$U_i = \sum_{j=1}^{s} n_{ij} u_{ij}$$

Then, the overall health outcomes U_{total} is computed as

$$U_{total} = \sum_{i=1}^{c} U_i$$

Cost may be computed analogously, replacing the U_i with corresponding cost estimates. Depending on the time horizon of treatment, but generally if over one year, discounting factors will also need to be incorporated into the summation. However, we omitted them here for simplicity.

In the alternative Monte Carlo simulation approach, patients are assumed to enter the model individually and transition through the states at random. Overall health outcome and cost values are then obtained by summing the respective individual patient values as the patient goes through the system.

5.4.5 Use in CEA

Once the values of health outcomes and costs are determined, CEAs may be performed using one of the techniques discussed in Section 5.2.2. While ICERs are commonly reported, they are not always readily interpretable to health care policy-makers for use in decision-making; for example, what represents a "good value" in an ICER may not be well-determined. Therefore, it may be advisable to also present results in other ways that facilitate decision-making, including providing the probability that a given intervention is less costly compared with a competitor subject to efficacy and resource constraints (Briggs et al., 2006). This is particularly relevant in budget impact analysis, that is, in evaluating the financial consequences of choosing a new intervention while taking into account resource constraints (Mauskopf et al., 2005). The cost-effectiveness probability approach discussed earlier in this chapter is relevant in this context.

5.4.6 Sensitivity Analysis

Since decision models are based on certain assumptions about the input parameters, it is important to perform analysis by varying relevant input values. In Markov models, for example, transition probabilities may be treatment-sequence-dependent, or may vary across population groups. The sensitivity analysis may include an evaluation of assumption about relevant model inputs (e.g., transition probabilities, health state utilities, resource use, cycles), or generalizability (e.g., patient characteristics, clinical setting). Such sensitivity analyses can help to determine the robustness and reliability of the results, and, hence, the level of confidence in the health care decision the analyses are intended to support. If the result is sensitive to variations in some of the parameters or assumptions, then appropriate adjustments should be made to the input variables, as well as to other aspects of the model.

In addition, the variability of the estimated quantities should also be quantified via probabilistic sensitivity analysis. This requires estimating distributions through either first-order simulation or second-order simulation. In the former, the variability of the estimates is obtained by simulating individual patient experience. In the latter, the parameters are assumed to have a probability distribution, and samples from these distributions are used to approximate the distributions of the estimator of interest; for example, see Briggs et al. (1994) and Nuijten (2004) for alternative approaches.

In some situations, the relative difference in expected health outcomes among alternative health care options may not be constant, but instead may depend on the values of the input parameters. More specifically, no single decision may be a preferred option for all values of the parameter. In this case, the expected value of the outcome as a function of the input parameter is plotted for each decision option, and the lines are inspected to see if they intersect, suggesting a threshold health state utility. See, for instance, Ryder et al. (2009) for an illustration of the approach using a one-way sensitivity analysis in which the utility for a clinical condition is systematically varied on a range of values.

In summary, while decision analysis has considerable importance in formulating health care policies, taking into account relevant health outcomes, it should be understood that inherent limitations can significantly impact health care resource utilization. A well-executed decision analysis, involving sensitivity analysis, may be time-consuming, and the process may also appear opaque to users of the results. Further, the availability and reliability of relevant input data may not be transparent, and there may often be a tendency to over-simplify complex health care issues for modeling expediency. Accordingly, it is critical that the analysis be executed judiciously and the results scrutinized with the utmost care.

5.4.7 CEA Example Using a Markov Model

In this section, we provide a simple cost-effectiveness model that illustrates how cost-effectiveness evaluations are performed in terms of treatment outcomes, costs, and utility weights, resulting in an ICER. For the illustration, a simple Markov model is used, based on artificial data. It is also demonstrated how sensitivity analysis may be performed.

Consider a simple three-period (cycle) model of a chronic disease with mild disease, severe disease, and death as health states, and two possible treatments, usual care and a new treatment. With usual care, a mild patient has a 60% probability of remaining mild, a 30% chance of progressing to severe, and a 10% chance of death, in each period. A severe patient has a 30% chance of improving to mild, a 40% chance of remaining severe, and a 30% chance of death in each period. Death is, of course, the absorbing (no-exit) state. This transition matrix is represented in Table 5.3. The corresponding matrix for the new treatment, which is expected to reduce disease worsening and mortality, is given in Table 5.4.

Assume that the new treatment is only effective for use in two periods (periods 1 and 2), affecting the outcomes for periods 2 and 3. The new treatment cost is $1,000 per period. Disease cost (i.e., other disease-related health care costs) per period is $100 for mild patients and $1,000 for severe

TABLE 5.3

Transition Matrix for Usual Care

Transition Probabilities between States	To		
From	*Mild*	*Severe*	*Dead*
Mild	0.6	0.3	0.1
Severe	0.3	0.4	0.3
Dead	0.0	0.0	1.0

TABLE 5.4

Transition Matrix for New Treatment

Transition Probabilities between States	To		
From	*Mild*	*Severe*	*Dead*
Mild	0.7	0.2	0.1
Severe	0.4	0.4	0.2
Dead	0.0	0.0	1.0

patients. Health state utilities are 0.9 for mild disease, 0.6 for severe disease, and 0 for death. For simplicity, assume that costs and outcomes after period 3 are not relevant for decision-making, and that discount rates are not necessary.

Suppose 1,000 patients enter period 1 while having mild disease. In one evaluation of the model, all patients receive usual care, transition to health states as shown in Table 5.3, experience utility (we will equate them to QALYs), and accrue disease costs for all three periods. In a second evaluation, all (alive) patients receive the new treatment in periods 1 and 2, and transition to health states as shown in Table 5.4, accruing treatment costs in the first two periods, and disease costs and QALYs in periods 1–3. The difference in total costs (including both disease and treatment costs) is then compared with the difference in total QALYs experienced in each state to calculate the ICER for the new treatment as compared with usual care.

For example, in the usual care arm, among 1,000 mild patients in period 1, 600 (1,000 × 0.6) will remain mild in period 2. Among those 600, 360 will remain mild in period 3, 180 will progress to severe in period 3, and 60 will die. Results for the usual care evaluation are shown in Table 5.5, with patients "moving" from left to right as they transition from one period to the next. Overall, 250 patients will die, 100 in period 2 and 150 in period 2. The total disease cost over three periods is $805,000 (with no new treatment costs in this evaluation), and patients experience 2,205 QALYs.

Similar calculations for patients with the new treatment are shown in Table 5.6. With the new treatment, 490 patients stay mild for both periods 2

TABLE 5.5

Results of Cost and Effectiveness Evaluation for Usual Care

					Usual Care								
Start	Period 1			Move to	Period 2			Move to	Period 3				
State	N	Costs ($1000)	QALYs	State	N	Costs ($1000)	QALYs	State	N	Costs ($1000)	QALYs	Totals	
Mild	1000	$100	900	Mild	600	$60	540	Mild	360	$36	324		
								Severe	180	$180,000	108		
								Dead	60	–	0		
				Severe	300	$300	180	Mild	90	$9	81		
								Severe	120	$120	72		
								Dead	90	–	0		
				Dead	100	–	0	Dead	100	–	0		
Total Disease Costs		$100				$360				$345		$715	
Total QALYs			900				720				585	2,205	

TABLE 5.6

Results of Cost and Effectiveness Evaluation for New Treatment

Start				New Treatment Move to				Move to				Totals
		Period 1				Period 2				Period 3		
State	N	Costs ($1000)	QALYs	State	N	Costs ($1000)	QALYs	State	N	Costs ($1000)	QALYs	
Mild	1000	$100	900	Mild	700	$70	630	Mild	490	$49	441	
								Severe	140	$140	84	
								Dead	70	–	0	
				Severe	200	$200	120	Mild	80	$8	72	
								Severe	80	$80	48	
								Dead	40	–	0	
				Dead	100	–	0	Dead	100	–	0	
Total Disease Costs		$100				$270				$277		$647
Total Treatment Costs		$1,000				$900						$1,900
Total QALYs			900				750				645	2,295

and 3, while only 210 (100 in period 2 and 110 in period 3) patients die. The total new treatment cost is $1,900,000, the disease cost is $647,000, and QALYs are 2,295.

The incremental costs of the new treatment can then be calculated as ($1,900,000+ $647,000 - $805,000) = $1,742,000, while the incremental QALYs due to the new treatment are (2,295–2,205) = 90. Dividing these two results yields an ICER of $19,356 per QALY saved. Hence, while the use of the new treatment would be more costly overall, it may still be considered a good value since its ICER is at a level that would be considered cost-effective in most developed countries (Eichler et al., 2004).

Simple sensitivity analyses could be performed by increasing the cost of the new treatment to $1,500; this change would not affect outcomes, disease costs, or QALYs, but would raise treatment costs to $2,850,000, and the ICER to $29,911 per QALY saved. If the utility of the severe state was lowered to 0.5, it would not affect costs, but would increase QALYs saved to 108—since the new treatment results in fewer severe patients—and reduce the ICER (at the $1,000 treatment cost) to $16,130. The reader may perform additional sensitivity analyses with varying relevant input values and model assumptions.

5.5 Conclusion

Health technology assessment is widely used in many parts of the world to guide decisions regarding the allocation of scarce resources to improve public health. The demands of scarce resources on the health care system and on the patient necessitate the assessment of the relative values of different health care interventions or treatments. Accordingly, several approaches have been proposed to facilitate health economic analysis, including CEA decision analysis. The reliability of results from such analyses, however, is heavily dependent on the validity of the underlying model assumptions, the degree to which the input variables are accurately determined, and the soundness of the theoretical basis for the statistical models. In this chapter, we provided a general overview of the commonly used measures and statistical approaches, with a particular emphasis on measures that need to be taken to mitigate the drawbacks of the approaches.

Statistical methodology plays an important role here, since the analysis of the data on cost and effectiveness is a critical aspect in this assessment. The choice of the criteria to be used for CEA and accurate statistical inference, based on the chosen criteria, are both crucial components of the data analysis. In our brief review, we have focused on both aspects, discussing traditional criteria and an accurate inference for them, and reviewing a new criterion (the CEP), and an accurate inference for it.

A key determining factor in any CEA endeavor is the availability of reliable data on cost and effectiveness. With regards to the latter, there has been extensive ongoing work, especially in the context of comparative effectiveness research in the United States, to address the evidentiary gap created either by the absence or inadequacy of RCT data. Nonetheless, the literature pertaining to CEA, while growing (Neumann et al., 2009), does not cover many possible treatment situations. This may in part be because of the fact that CEA is not widely used in the US, as compared with other industrialized countries, in health care decision-making (Bryan et al., 2009). It is, however, hoped that with the ever-increasing cost of health care, researchers involved in CEA will continue to take advantage of methodological developments to enhance the evidentiary value of CEA results, so as to promote health care delivery and utilization.

Personalized medicine is now an emerging field, due to the diversified nature of various diseases, and the variations in patient characteristics. Nevertheless, the literature for CEA in personalized medicine is very limited; see Ioannidis and Garber (2011). In a recent book, Fragoulakis et al. (2015) brought up a number of issues in the context of CEA for personalized medicine; the topic appears to be in its initial stages only, and the authors list a number of problems that require methodological development. In particular, while assessing the effectiveness of several competing health interventions, the authors consider using a "linear combination" of treatments as an option; for example, if there are two treatments, half of the patients receive one treatment, and the remaining half receive the other treatment. Statistical methodology for CEA in such mixed scenarios, and more generally in the context of personalized medicine, is a topic that needs investigation.

Acknowledgment

Special thanks to Yang Chen for help with typesetting.

References

Alemayehu, D. 2011. Assessing exchangeability in indirect and mixed treatment comparisons. *J Comp Eff Res* 1:51–55.

Baio, G. 2012. *Bayesian Methods in Health Economics*. Boca Raton, FL: Chapman & Hall/CRC.

Bang, H. and H. Zhao. 2012. Average cost-effectiveness ratio with censored data. *J Biopharm Stat* 22:401–415.

Barendregt, J.J. 2009. The half-cycle correction: Banish rather than explain It. *Med Decis Making* 29:500–502.

Barton, G.R., Briggs, A.H., and E. Fenwick. 2008. Optimal cost-effectiveness decisions: The role of the cost-effectiveness acceptability curve (CEAC), the cost-effectiveness acceptability frontier (CEAF), and the expected value of perfection information (EVPI). *Value Health* 11:886–897.

Bebu, I., Luta, G., Mathew, T. et al. 2016a. Parametric cost-effectiveness inference with skewed data. *Comput Stat Data Anal* 94:210–220.

Bebu, I. and T. Mathew. 2008. Comparing the means and variances of a bivariate lognormal distribution. *Stat Med* 27:2684–2696.

Bebu, I., Mathew, T., and J.M Lachin. 2016b. Probabilistic measures of cost-effectiveness. *Stat Med* 25:3976–3986.

Beck, J.R. and S.G. Pauker. 1983. The Markov process in medical prognosis. *Med Decis Making* 3:419–458.

Berger, M.L., Mamdani, M., Atkins, D. et al. 2009. Good research practices for comparative effectiveness research: Defining, reporting and interpreting nonrandomized studies of treatment effects using secondary data sources: The ISPOR good research practices for retrospective database analysis task force report— Part I. *Value Health* 12:1044–1052.

Briggs, A. and M. Sculpher. 1998. An introduction to Markov modelling for economic evaluation. Pharmacoeconomics 13:397 -409.

Briggs, A., Sculpher, M., and M. Buxton. 1994. Uncertainty in the economic evaluation of health care technologies: The role of sensitivity analysis. Health Econ 3:95 -104.

Briggs, A., Sculpher, M., and K. Claxton. 2006. Decision Modeling for Health Economic Evaluation. New York, NY: Oxford University Press.

Briggs, A.H. 2010. Transportability of comparative effectiveness and cost-effectiveness between countries, Value Health 13:S22–S25.

Briggs, A.H. and P. Fenn. 1997. Trying to do better than average: A commentary on statistical inference for cost-effectiveness ratios. *Health Econ* 6:491–495.

Briggs, A.H., Mooney, C.Z., and D.E. Wonderling. 1999. Constructing confidence intervals for cost-effectiveness ratios: An evaluation of parametric and non-parametric techniques using Monte Carlo simulation. *Stat Med* 18:3245–3262.

Bryan, S., Sofaer, S., Siegelberg, T. et al. 2009. Has the time come for cost-effectiveness analysis in U.S. health care? *Health Econ Policy Law* 4:425–443.

Bucher, H.C., Guyatt, G.H., Griffith, L.E. et al. 1997. The results of direct and indirect treatment comparisons in meta-analysis of randomized controlled trials. *J Clin Epidemiol* 50:683–691.

Burstrom, K., Johannesson, M., Diderichsen, F. et al. 2006. A comparison of individual and social time-trade-off values for health states in the general population. *Health Policy* 76:359–370.

Caldwell, D.M., Ades, A.E., and J.P.T. Higgins. 2005. Simultaneous comparison of multiple treatments: Combining direct and indirect evidence. *BMJ* 331:897–900.

Chaudhary, A.M. and C.S. Stearns. 1996. Estimating confidence intervals for cost-effectiveness ratios: An example from a randomized trial. *Stat Med* 15:1447–1458.

Cox, E., Marin, B.C., Van Staa, T. et al. 2009. Good research practices for comparative effectiveness research: Approaches to mitigate bias and confounding in the design of nonrandomized studies of treatment effects using secondary data sources: The ISPOR Good research practices for retrospective database analysis task force report – Part II. *Value Health* 12:1053–1061.

Coyle, D., Coyle, K., Cameron, C. et al. 2014. Benefits from incorporating network meta-analysis within stratified cost-effectiveness analysis. https://www.ispor.org/awards/19meet/MO2.pdf (accessed May 16, 2017).

Deb, P. and P.K. Trivedi. 2002. The structure of demand for health care: Latent class versus two-part models. *J Health Econ* 21:601–625.

Dias, S., Welton, N.J., Caldwell, D.M. et al. 2010. Checking consistency in mixed treatment comparison meta-analysis. *Stat Med* 29:932–944.

Dias, S., Welton, N.J, Sutton, A.J. et al. 2013. Evidence synthesis for decision making 4: Inconsistency in networks of evidence based on randomized controlled trials. *Med Decis Making* 33:641–656.

Dolan, P., Gudex, C., Kind, P. et al. 1996. Valuing health states: A comparison of methods. *J Health Econ* 15:209–231.

Doubilet, P., Weinstein, M.C., and B.J. McNeil. 1986. Use and misuse of the term "cost effective" in medicine. *N Engl J Med* 314:253–256.

Drummond, M., Barbieri, M., Cook, J. et al. 2009. Transferability of economic evaluations across jurisdictions: ISPOR good research practices task force report. *Value Health* 12:409–418.

Drummond, M.F., Richardson, W.S., O'Brien, B.J. et al. 1997. Users' guides to the medical literature. XIII. How to use an article on economic analysis of clinical practice. A. Are the results of the study valid? Evidence-Based Medicine Working Group. *JAMA* 277:1552–1557.

Duan, N. 1983. Smearing estimate: A nonparametric retransformation method. *J Am Stat Assoc* 78:605–610.

Duan, N., Manning, W.G. Jr., Morris, C.N. et al. 1983. A comparison of alternative models of the demand for medical care. *J Bus Econ Stat* 1:115–126.

Eichler, H.G., Kong, S.X., Gerth, W.C. et al. 2004. Use of cost-effectiveness analysis in health-care resource allocation decision-making: How are cost-effectiveness thresholds expected to emerge? *Value Health* 7:518–528.

Fan, M.Y. and X.H. Zhou. 2007. A simulation study to compare methods for constructing confidence intervals for the incremental cost-effectiveness ratio. *Health Serv Outcomes Res Methodol* 7:57–77.

Fenwick, E. and S. Byford. 2005. A guide to cost-effectiveness acceptability curves. *Br J Psychiatry* 187:106–108.

Fenwick, E., Claxton, K., and M.J. Sculpher. 2001. Representing uncertainty: The role of cost-effectiveness acceptability curves. *Health Econ* 10:779–787.

Fenwick, E., O'Brien, B.J., and A. Briggs. 2004. Cost-effectiveness acceptability curves – facts, fallacies and frequently asked questions. *Health Econ* 13:405–415.

Fragoulakis, V., Mitropoulou, C., and M.S. Williams. 2015. *Economic Evaluation in Genomic Medicine*. Amsterdam, The Netherlands Elsevier.

Faria, R., Alava, M.H., Manca, A. et al. 2015. *NICE DSU Technical Support Document 17: The use of observational data to inform estimates of treatment effectiveness for Technology Appraisal: Methods for comparative individual patient data*. http://www.nicedsu.org.uk (accessed July 7, 2016).

Foucher, Y., Mathieu, E., Saint-Pierre, P. et al. 2006. A semi-Markov model based on Generalized Weibull distribution with an illustration for HIV disease. *Biom J* 47:825–833.

Gafni, A. 1994. The standard gamble method: What is being measured and how it is interpreted. *Health Serv Res* 29:207–224.

Gamage, J., Mathew, T., and S. Weerahandi. 2004. Generalized p-values and generalized confidence regions for the multivariate Behrens Fisher problem and MANOVA. *J Multivar Anal* 88:177–189.

Gomes, M., Grieve, R., Nixon, R. et al. 2012a. Methods for covariate adjustment in cost-effectiveness analyses that use cluster randomized trials. *Health Econ* 21:1101–1118.

Gomes, M., Grieve, R., Nixon, R. et al. 2012b. Statistical methods for cost-effectiveness analyses that use data from cluster randomized trials: A systematic review and checklist for critical appraisal. *Med Decis Making* 32:209–220.

Hawkins, N., Sculpher, M., and D. Epstein. 2005. Cost-effectiveness analysis of treatments for chronic disease: Using R to incorporate time dependency of treatment response. *Med Decis Making* 25:511–519.

Hershey, J.C., Asch, D.A., Jepson, C. et al. 2003. Incremental and average cost-effectiveness ratios: Will physicians make a distinction? *Risk Anal* 23:81–89.

Hoaglin, D.C., Hawkins, N., Jansen, J.P. et al. 2011. Conducting indirect-treatment-Comparison and network-meta-analysis studies: Report of the ISPOR task force on indirect treatment comparisons good research practices—Part 2. *Value Health* 14:429–437.

Hoch, J.S. and C.S. Dewa. 2008. A clinician's guide to correct cost-effectiveness analysis: Think incremental not average. *Can J Psychiatry* 53:267–274.

Hunink, M., Glasziou, P., Siegel, J. et al. 2001. *Decision Making in Health and Medicine: Integrating Evidence and Values.* Cambridge, UK: Cambridge University Press.

Ioannidis, J.P.A. and A.M. Garber. 2011. Individualized cost-effectiveness analysis. *PLoS Med* 8(7):e1001058. https://doi.org/10.1371/journal.pmed.1001058 (accessed May 16, 2017).

Jiang, G., Wu, J., and G.R. Williams. 2000. Fieller's interval and the Bootstrap-Fieller interval for the incremental cost-effectiveness ratio. *Health Serv Outcomes Res Methodol* 1:291–303.

Johnson, M.L., Crown, W., Martin, B.C. et al. 2009. Good research practices for comparative effectiveness research: Analytic methods to improve causal inference from nonrandomized studies of treatment effects using secondary data sources: The ISPOR good research practices for retrospective database analysis task force report—Part III. *Value Health* 12:1062–1073.

Karnon, J. and J. Brown. 1998. Selecting a decision model for economic evaluation: A case study and review. *Health Care Manag Sci* 1:133–140.

Khoo, A.L., Zhou, H.J., Teng, M. et al. 2015. Network meta-analysis and cost-effectiveness analysis of new generation antidepressants. *CNS Drugs* 29:695–712.

Laska, E.M., Meisner, M., and C. Siegel. 1997a. Statistical inference for cost-effectiveness ratios. *Health Econ* 6:229–242.

Laska, E.M., Meisner, M., and C. Siegel. 1997b. The usefulness of average cost-effectiveness ratios. *Health Econ* 6:497–504.

Lu, G. and A.E. Ades. 2004. Combination of direct and indirect evidence in mixed treatment comparisons. *Stat Med* 23:3105–3124.

Luce, B. R., Manning, W. G., Siegel, J. E., and Lipscomb, J. 1996. Estimating costs in cost-effectiveness analysis. In Cost-Effectiveness in Health and Medicine (Gold, M., Siegel, J. E., Russell, L. B and Weinstein, M. C., Editors). New York: Oxford University Press, pp. 176-213.

Lumley, T. 2002. Network meta-analysis for indirect treatment comparisons. *Stat Med* 21:2313–2324.

Madigan, D., Ryan, P.B., Schuemie, M. et al. 2013a. Evaluating the impact of database heterogeneity on observational study results. *Am J Epidemiol* 178:645–651.

Madigan, D., Ryan, P.B., Schuemie, M. et al. 2013b. Does design matter? Systematic evaluation of the impact of analytical choices on effect estimates in observational studies. *Ther Adv Drug Saf* 4:53–62.

Manning, W.G. 1998. The logged dependent variable, heteroscedasticity, and the retransformation problem. *J Health Econ* 17:283–295.

Manning, W.G., Basu, A., and J. Mullahy. 2005. Generalized modeling approaches to risk adjustment of skewed outcomes data. *J Health Econ* 24:465–488.

Manning, W.G. and J. Mullahy. 2001. Estimating log models: To transform or not to transform? *J Health Econ* 20:461–494.

Mauskopf, J.A., Earnshaw, S., and C.D. Mullins. 2005. Budget impact analysis: Review of the state of the art. *Future Drugs Ltd* 5:65–79.

McGhan, W.F., Al, M., Doshi, J.A. et al. 2009. The ISPOR good practices for quality improvement of cost-effectiveness research task force report. *Value Health* 12:1086–1099.

Mihaylova, B., Briggs, A., O'Hagan, A. et al. 2011. Review of statistical methods for analyzing health care resources and costs. *Health Econ* 20:897–916.

Mullahy, J. 1998. Much ado about two: Reconsidering retransformation and the two-part model in health econometrics. *J Health Econ* 17:247–281.

Neumann, P.J., Fang, C.H., and J.T. Cohen. 2009. 30 years of pharmaceutical cost-utility analyses: Growth, diversity and methodological improvement. *Pharmacoeconomics* 27:861–872.

Neupane, B., Richer, D., Bonner, A.J. et al. 2014. Network meta-analysis using R: A review of currently available automated packages. *PLOS ONE* 9(12):e115065. https://doi.org/10.1371/journal.pone.0115065 (accessed May 17, 2017).

Nixon, R.M., Wonderling, D., and R. Grieve. 2005. How to estimate cost-effectiveness acceptability curves, confidence ellipses and incremental net benefits alongside randomized controlled trials. *Technical Report of the Medical Research Council Biostatistics Unit*, Cambridge.

Nixon, R.M., Wonderling, D., and R.D. Grieve. 2010. Non-parametric methods for cost-effectiveness analysis: The central limit theorem and the bootstrap compared. *Health Econ* 9:316–333.

Nuijten, M.J. 2004. Incorporation of statistical uncertainty in health economic modelling studies using second-order Monte Carlo simulations. *Pharmacoeconomics* 22:759–769.

O'Brien, B.J., Drummond, M.F., Labelle, R.J. et al. 1994. In search of power and significance: Issues in the design and analysis of stochastic cost effectiveness studies in health care. *Med Care* 32:150–163.

Polsky, D., Glick, H.A, Willke, R. et al. 1997. Confidence intervals for cost-effectiveness ratios: A comparison of four methods. *Health Econ* 6:243–252.

Rabin, R. and F. de Charro. 2001. EQ-5D: A measure of health status from the EuroQol Group. *Ann Med* 33:7–43.

Ramsey, S.D., Willke, R.J., Glick, H.A. et al. 2015. Cost-effectiveness analysis alongside clinical trials II: An ISPOR good research practices task force report. *Value Health* 18:161–172.

Revicki, D.A., Brown, R.E., Palmer, W. et al. 1995. Modelling the cost effectiveness of antidepressant treatment in primary care. *Pharmacoeconomics* 8:524–540.

Rosenbaum, P.R. and D.B. Rubin. 1983. The central role of the propensity score in observational studies for causal effects. *Biometrika* 70:41–55.

Ryder, H.F., McDonough, C., Tosteson, A.N.A. et al. 2009. Decision analysis and cost-effectiveness analysis. *Semin Spine Surg* 21:216–222.

Salas, M., Hofman, A., and B.H. Stricker. 1999. Confounding by indication: An example of variation in the use of epidemiologic terminology. *Am J Epidemiol* 149:981–983.

Signorovitch, J.E., Wu, E.Q., Yu, A.P. et al. 2010. Comparative effectiveness without head-to-head trials: A method for matching-adjusted indirect comparisons applied to psoriasis treatment with adalimumab or etanercept. *Pharmacoeconomics* 28:935–945.

Thompson, S.G. and R.M. Nixon. 2005. Methods for incorporating covariate adjustment, subgroup analysis and between-centre differences into cost-effectiveness evaluations. *Health Econ* 14:1217–1229.

Torrance, G.W. 1976. Social preferences for health states: An empirical evaluation of three measurement techniques. *Soc Econ Plann Sci* 10:128–136.

Tosh, J. and A. Wailoo. 2008. Review of software for decision modelling. *NICE Decision Support Unit Report.* http://www.nicedsu.org.uk/PDFs%20of%20reports/softwarereport-final.pdf (accessed May 15, 2017).

von Neumann, J. and O. Morgenstern. 1953. *Theory of Games and Economic Behavior,* 3rd edition. New York, NY: Wiley.

Weerahandi, S. 1993. Generalized confidence intervals. *J Am Stat Assoc* 88:899–905

Weinstein, M.C. and H.V. Fineberg. 1980. *Clinical Decision Analysis.* Philadelphia, PA: W.B. Saunders.

Weinstein, M.C., Siegel, J.E., Gold, M.R. et al. 1996. Recommendations of the panel on cost-effectiveness in health and medicine. *JAMA* 276:1253–1258.

Weinstein, M.C. and W.B. Stason. 1977. Foundations of cost-effectiveness analysis for health and medical practices. *N Engl J Med* 296:716–721.

Willan, A.R. and A.H. Briggs. 2006. *Statistical Analysis of Cost-Effectiveness Data.* Chichester, UK: Wiley.

Willan, A.R. and B.J. O'Brien. 1996. Confidence intervals for cost-effectiveness ratios: An application of Fieller's theorem. *Health Econ* 5:297–305.

Zhao, Y.J., Khoo, A.L., Tan, G. et al. 2016. Network meta-analysis and pharmacoeconomic evaluation of fluconazole, itraconazole, posaconazole, and voriconazole in invasive fungal infection prophylaxis. *Antimicrob Agents Chemother* 60:376–386.

6

Analysis of Aggregate Data

Demissie Alemayehu, Andrew G. Bushmakin,
and Joseph C. Cappelleri

CONTENTS

6.1 Introduction 124
6.2 Traditional Meta-Analysis 125
    6.2.1 Frequentist Framework 125
    6.2.2 Bayesian Framework 126
    6.2.3 Cumulative Meta-Analysis 127
6.3 NMA 128
    6.3.1 Fixed Effects Model 130
    6.3.2 Random Effects Model 131
6.4 Model Validation 131
    6.4.1 Traditional Meta-Analysis 131
    6.4.2 NMA 132
    6.4.3 Publication Bias 134
6.5 Meta-Regression 135
6.6 Best Practices for the Conduct and Reporting of
    Meta-Analysis 136
    6.6.1 Cochrane Collaboration's Risk of Bias Assessment:
        A Tool for Evaluating the Quality of Individual
        Randomized Trials 137
    6.6.2 PRISMA Suite, MOOSE, and MAER-Net: Tools for
        Reporting 137
    6.6.3 AMSTAR and GRADE: Tools to Appraise the
        Quality of Systematic Reviews 139
    6.6.4 Tools to Appraise the Quality of Systematic
        Reviews 139
6.7 Random Effects Meta-Analysis: A Simulated
    Example Using SAS® 140
6.8 Discussion 143
References 144

6.1 Introduction

In this chapter, we consider statistical issues that arise when the analysis involves synthesizing information based on summary statistics from independent sources on the same topic of interest. Such analyses may be executed either to generate a new hypothesis (retrospective analysis) or to confirm an existing one (prospective analysis). A distinction is often made between systematic reviews—which are intended to minimize selection bias through a formal process involving prespecification of study objectives, candidate studies, and analytical approaches—and traditional narrative reviews, which do not follow a systematic search of the literature (Uman, 2011).

Meta-analysis may be defined as the statistical analysis of data from multiple studies. Strictly speaking, it requires only two studies, and need not be based on a systematic review. However, meta-analysis in medical research typically identifies data systematically from multiple studies, summarizes results, and quantitatively evaluates sources of heterogeneity and bias. A systematic literature review encompasses an explicit and detailed description of how a review was conducted. Some have characterized meta-analysis broadly to include a literature systematic review, as well as a statistical analysis of such data. In this sense, meta-analysis has been also referred to as a quantitative systematic review or an overview (Borenstein et al., 2009; Cappelleri et al., 2010; Egger et al., 2001).

In the traditional meta-analysis framework, the objective of the analysis may consist of combining information from trials in which the same treatment groups were studied and compared (Schmid et al., 1991). In this case, meta-analysis may be used to address uncertainty and heterogeneity when results of studies disagree, to increase statistical power for primary outcomes and subgroups, to improve the precision of estimates of treatment effect, or to lead to new knowledge and the formulation of new questions (Sacks et al., 1987). Meta-analysis is also a powerful tool for exploring sources of heterogeneity and bias in clinical research. In addition, there are also instances where it may be essential or at least desirable to draw inference about the relative risks and benefits of alternative treatment options that have not been studied in head-to-head randomized controlled trials (RCTs) (Bucher et al., 1997). The goal in this situation is to apply techniques that maximize and capitalize on the benefits of the randomization in the original trials.

While the importance of combining information from independent sources on the same topic is widely recognized, there are also fundamental methodological and conceptual issues that need to be addressed to ensure the reliability of the results in health care decision-making (Egger and Smith, 1995; Thompson and Pocock, 1991; Senn et al., 2013; Song et al., 2003). Accordingly, numerous guidelines have been proposed to strengthen the value of results from such analyses, in order to help decision-makers assess the robustness of the reported findings (see, e.g., Baker et al., 2016; Hoaglin et al., 2011; Moher et al., 2009). Further, there have been funding opportunities to support research and infrastructure development in this area by various groups, including the Patient-Centered Outcomes Research Institute in the United States (Mayo-Wilson, 2015).

In the following sections, we provide a high-level overview of the statistical issues associated with both traditional meta-analysis (Section 6.2) and the more recent techniques used in network meta-analysis (NMA) (Section 6.3). In particular, a formal expression of the underlying models is provided, accompanied by an account of the relevant assumptions and measures needed to mitigate the impacts of deviations from those assumptions. In addition, examples are provided to illustrate pertinent topics. We also cover model validation (Section 6.4) and meta-regression (Section 6.5). Section 6.6 provides a brief review of the recent literature on best practices for the conduct and reporting of meta-analyses. Section 6.7 presents a computer simulation of random effects meta-analysis. Section 6.8 provides a brief discussion, in the form of a concluding commentary.

6.2 Traditional Meta-Analysis

6.2.1 Frequentist Framework

Consider combining the estimates of treatment effects on the same effects metric from s studies, which are selected according to some prespecified criteria. For the ith study, $(i = 1, ..., s)$, denote the statistic and its associated estimator of variance by $\hat{\theta}_i$ and $\hat{v}_i$, respectively. Typical examples of $\hat{\theta}_i$ include suitably defined functions of such effect measures as mean differences, odds ratios, risk ratios, or risk differences, which are assumed to be approximately normally distributed (Borenstein et al., 2009; Cooper et al., 2009a).

In practice, the analysis is performed using either a fixed effects or a random effects model in a frequentist or Bayesian framework (Borenstein et al., 2009; DerSimonian and Laird, 1986). Under the frequentist paradigm, which traditionally tends to be more popular than the Bayesian paradigm for direct comparisons of treatments, the fixed effects model assumes that the study-specific statistics (estimated treatment effects) have a common underlying mean, θ, which is estimated as a weighted average of the $\hat{\theta}_i$s as follows:

$$\hat{\theta} = \frac{\sum_{i=1}^{s} w_i \hat{\theta}_i}{\sum_{i=1}^{s} w_i}$$

where, typically, $w_i = 1/\hat{v}_i$ (i.e., weighted by the inverse of the variance estimator of the estimated treatment effect in the ith study). A consequence of this formulation is that larger studies tend to get more weight, whereby the contribution of smaller studies is dampened. In addition, the setting assumes that the studies are essentially homogenous with regards to relevant attributes, including study conduct, and study and patient characteristics, as well as outcome measures. In statistical terminology, this means that there is no treatment-by-study interaction or effects modification. The latter is usually tested using the Q statistic, given by $Q = \sum_{i=1}^{s} w_i (\theta_i - \hat{\theta})^2$. Under the null hypothesis of homogeneity, Q has an approximate χ^2 distribution with $s-1$ degrees of freedom.

In contrast, the random effects model takes into account between-study variability. More specifically, $\hat{\theta}_i$ is assumed to have mean θ_i and variance v_i, where the θ_is are in turn assumed to have a common underlying mean θ and variance σ_B^2. An estimate of θ is given by

$$\hat{\theta} = \frac{\sum_{i=1}^{s} w_i^* \hat{\theta}_i}{\sum_{i=1}^{s} w_i^*}$$

with $w_i^* = 1/(v_i + \sigma_B^2)$. In practice, the unknown variance quantities (v_i, σ_B^2) are replaced by their corresponding sample estimates. It should be noted that at least some heterogeneity in treatment effects across studies is a given. In the presence of substantial heterogeneity, however, the interpretation of results based on the random effects formulation requires caution. Further, when the number of studies is small, the between-study variance σ_B^2 cannot be estimated reliably, leading to an incorrect homogeneity conclusion (Hardy and Thompson, 1996).

A multitude of traditional meta-analysis publications has appeared in the literature over the last few decades. Two examples that illustrate the use of frequentist meta-analysis are Tan et al. (2015) and Whiting et al. (2015). The first of these was conducted to systematically review studies about the use of dual therapy (clopidogrel and aspirin) in comparison with monotherapy (aspirin alone) for stroke prevention, and included five randomized studies involving 24,084 patients. The second example was aimed at evaluating the benefits and adverse events of the use of cannabinoids relative to a placebo, and consisted of a total of 79 trials, with 6,462 participants.

6.2.2 Bayesian Framework

Meta-analysis in a Bayesian framework is now increasingly used, typically mirroring the frequentist setup. Presumed advantages of the approach include the ability to incorporate prior information, the flexibility of the model to tackle complex scenarios, the ability to derive a predictive distribution for the effect expected in a new trial, and, perhaps most attractively, the ability to enhance the interpretation and meaning of inferences through probabilistic statements regarding quantities of interest (e.g., the probability of better median survival for patients receiving drug A as compared with patients receiving drug B) (Sutton and Abrams, 2001).

Examples of common priors for the treatment effect include a normal distribution with a null mean (no difference) and a large variance if a noninformative prior is used, and a normal distribution with a nonnull mean and a smaller variance if an informative prior is used. Although a uniform prior can be specified for σ_B^2, often alternative specifications are considered. For meta-analyses that include small or moderate numbers of studies, there will likely be little information regarding the estimation of σ_B^2 and therefore the prior can be influential in the analyses. The Jeffreys prior, which is noninformative, is frequently considered. Another possibility is to use a normal distribution truncated at zero and with large variance to characterize a suitably vague prior. Alternatively, one may use the inverse gamma distribution, which has the flexibility to incorporate several ranges of beliefs and thereby allows for a varying amount of informative (prior) information.

The use of Bayesian meta-analyses for traditional pairwise comparisons has increased over the years, in proportion to the availability of software for implementation. One example of a traditional Bayesian meta-analysis involved nine trials with 19,569 patients in the study of secondary prevention in elderly patients (Afilalo et al., 2008). In another instance, a Bayesian meta-analysis was performed to compare the benefits of steroids alone versus steroids plus antivirals for the treatment of Bell's palsy (Quant et al., 2009). The informative prior for treatment effect was taken from a previous study and based on the normal distribution; the inverse gamma distribution on between-study variance was specified.

6.2.3 Cumulative Meta-Analysis

Cumulative meta-analysis is a method of updating previous meta-analyses in light of the appearance of new studies (Lau et al., 1992). The method allows for the monitoring of developing trends of therapeutic efficacy. When performed routinely, the earliest time at which statistical significance is reached (by whatever criterion is chosen) can be identified.

More generally, cumulative analysis involves combining information from studies included in a sequential manner according to some ordered metric. In addition to the date of publication, the order in which studies are added may be a function of the effect size, trial quality, proportion of events in the control group (as an indicator of baseline risk in a study or sample), or other study or patient covariates of interest.

Cumulative meta-analysis inherently resides within a Bayesian framework (Lau et al., 1995). The cumulative aspect of the meta-analysis and its graphical representation exemplified in its figures are calculated, and can be interpreted through the Bayesian paradigm as new data to continually update the posterior distribution. The last posterior distribution becomes the next prior distribution. Thus, by inspecting the accumulating information, researchers can be informed so as to answer important decisions regarding the medical question(s) of interest. For example, with accumulating data, there may be sufficient evidence to make the need for additional studies superfluous (Clarke et al., 2014). Further, such data may also suggest the need to focus on certain patient populations that require additional scrutiny.

The first placebo-controlled randomized study of a thrombolytic drug in the treatment of acute myocardial infarction was reported in 1959. Over the subsequent years, 62 more placebo-controlled trials were reported, involving an additional 45,000 patients. Interestingly, a cumulative meta-analysis could have revealed that, by 1973, after eight studies included 2,432 patients, a statistically significant reduction in overall mortality was evident in favor of the use of thrombolytic drugs relative to a placebo (Antman et al., 1992). The Food and Drug Administration (FDA) did not approve a thrombolytic drug for this indication until 1988. Compared with the results of the cumulative meta-analysis, expert opinions lagged substantially in recognizing the efficacy of the treatment, and did not begin to recommend it until around the time of the FDA approval.

It is noted that one needs to consider the potential for false positive rates when performing standard significance tests in the context of cumulative meta-analysis (Berkey et al., 1996). Application of the law of iterated logarithm has been recommended in order to control the type 1 error (level of significance) in cumulative meta-analysis of binary and continuous outcomes (Hu et al., 2007; Lan et al., 2003). Some authors (see, e.g., Higgins et al., 2011a) propose the use of sequential methods in a semi-Bayesian framework. However, there does not seem to be a consensus on the applicability of sequential methods in cumulative meta-analysis.

6.3 NMA

An NMA is performed when interest lies in assessing the relative risks or benefits of alternative treatment options that have not been studied in the same RCT. If the available evidence consists of a network of multiple RCTs involving treatments compared directly, indirectly, or both, it can be synthesized by means of the so-called NMA (Lumley, 2002). Applications of NMAs in the published literature are becoming more and more common (Donegan et al., 2010, 2013).

The simplest form of an NMA, referred to as an indirect treatment comparison, is applied when only two treatments are being compared indirectly. Consider, for example, a situation where there is interest in evaluating the comparative effectiveness of two treatments, A and C, each of which has been studied separately against a common comparator, B (the placebo), in two separate RCTs. Let d_{AB} and d_{CB} denote the respective estimated effects in the two trials. Common examples of d_{AB} and d_{CB} include suitable functions of the usual efficacy measures, such as odds ratios, risk differences, or mean differences. In a seminal paper, Bucher et al. (1997) proposed estimating the effects of A relative to C indirectly by

$$d_{AC} = d_{AB} - d_{CB}$$

In contrast to a naïve-approach based on the direct treatment comparisons involving the estimates of effects of A and C only (without considering the effect of B), a flawed approach that does not adjust for potential confounders, the proposed technique preserves some of the benefits of randomization, as A is compared with B within the same RCT, and C is compared with B within the same RCT. The estimated standard error may be computed as

$$SE(d_{AC}) = \sqrt{\mathrm{var}(d_{AB}) + \mathrm{var}(d_{CB})}$$

Under the usual assumption of asymptotic normality, confidence intervals and test criteria may be constructed in the usual manner.

The above formulation, however, cannot handle more complex situations, especially when multiple pairs of treatments are involved and when both direct and indirect comparative evidence are present. NMA or mixed treatment comparison techniques are now available for indirect comparisons of more than two treatments (Lu and Ades, 2004; Wells et al., 2009). NMA can be performed with fixed or random effects models (Jansen et al., 2011; Hoaglin et al., 2011). Analyses based on these models could use either frequentist or Bayesian methods. In the following, fixed effects and random effects formulations are highlighted, along with frequentist and Bayesian versions. In the pertinent literature, Bayesian techniques appear to be more commonly applied, partly because the methods have computationally progressed to address the complexity imposed by networks of evidence.

As an example, a NMA was undertaken to assess the effects of 12 new-generation antidepressants on major depression (Cipriani et al., 2009). Based on a systematic review of 117 randomized controlled trials (25,928 participants), anti-depressants were quantified, compared, and ranked with respect to the proportion of patients who responded to the allocated treatment (efficacy) and, separately, the proportion who dropped out of the allocated treatment (acceptability). Clinically important differences were noted between commonly prescribed antidepressants with respect to both efficacy and acceptability.

6.3.1 Fixed Effects Model

With a fixed effects model, it is assumed that there is no variation in relative treatment effects across studies for a particular pairwise comparison. Observed differences for a particular comparison among study results are solely due to chance. For any given treatment comparison in a fixed effects model, the following question arises: "What is the true treatment effect?"

When the evidence network consists of multiple pairwise comparisons (i.e., AB trials, AC trials, BC trials, and so on), the set of comparators usually varies among studies, complicating the notation. One approach labels the treatments A, B, C, and so on, and uses A for the primary reference treatment in the analysis. In each study, it designates one treatment, b, as the base treatment. The labels can be assigned to treatments in the network in such a way that the base treatments follow A (i.e., B, C, and so on), and the nonbase treatments follow all the base treatments in the alphabet. In the various models, "after" refers to this alphabetical ordering. The general frequentist fixed effects model for NMA can then be specified as follows (Hoaglin et al., 2011):

$$\eta_{jk} = \begin{cases} \mu_{jb} & b = A,B,C, \quad \text{if } k = b \\ \mu_{jb} + d_{bk} = \mu_{jb} + d_{Ak} - d_{Ab} & k = B,C,D, \quad \text{if } k \text{ is "after" } b \end{cases}$$

where

η_{jk} reflects the underlying outcome for treatment k in study j,

μ_{jb} is the outcome for treatment b in study j, and

d_{bk} is the fixed effect of treatment k relative to treatment b.

The d_{bk} are identified by expressing them in terms of effects relative to treatment A as follows: $d_{bk} = d_{Ak} - d_{Ab}$ with $d_{AA} = 0$. For the underlying effects, this relation is a statement of consistency: the "direct" effect d_{bk} and the "indirect" effect $d_{Ak} - d_{Ab}$ are equal.

The corresponding general Bayesian fixed effects model would place a prior distribution on d_{Ak} whether the model is a fixed effects or a random effects, Bayesian methods combine the likelihood (roughly, the probability of the data as a function of the parameters) with a prior probability distribution (which reflects a prior belief about the possible values of those parameters) to obtain a posterior probability distribution of the parameters (Hoaglin et al., 2011; Sutton and Abrams, 2001). The posterior probabilities provide a straightforward way to make predictions, and the prior distribution can incorporate various sources of uncertainty. For parameters such as treatment effects, the customary prior distributions are noninformative. The assumption that, before seeing the data, all values of the parameter are equally likely helps to minimize the influence of the prior distribution on the posterior results. However, when information on the parameter is available (e.g., from observational studies, or from a previous analysis), the prior distribution provides a natural way to incorporate it.

6.3.2 Random Effects Model

If there is heterogeneity and, hence, variation across trials with respect to true (or underlying) relative treatment effects for a particular pairwise comparison, random effects models are appropriate (Jansen et al., 2011). A random effects model approach typically assumes that the trial-specific relative effects can be described as a sample from a normal distribution, whose standard deviation reflects the heterogeneity. With a random effects model for an NMA, the variance reflecting heterogeneity is often assumed to be constant for all pairwise comparisons.

As an extension of the frequentist fixed effects model, the frequentist random effects model replaces d_{bk} with δ_{jbk}, the trial-specific effect of treatment k relative to treatment b (Hoaglin et al., 2011). These trial-specific effects are drawn from a random-effects distribution: $\delta_{jbk} \sim N\left(d_{bk}, \sigma^2\right)$. Again, the d_{bk} are identified by expressing them in terms of the primary reference treatment, A. This model assumes, as stated in the prior paragraph, the same random-effect variance σ^2 for all treatment comparisons, but the constraint can be relaxed. (A fixed effects model results if $\sigma^2 = 0$.) Hence

$$\eta_{jk} = \begin{cases} \mu_{jb} & b = A,B,C, \quad \text{if } k = b \\ \mu_{jb} + \delta_{jbk} & k = B,C,D, \quad \text{if } k \text{ is "after" } b \end{cases}$$

$$\delta_{jbk} \sim N\left(d_{bk}, \sigma^2\right) = N\left(d_{Ak} - d_{Ab}, \sigma^2\right)$$

The corresponding random effects Bayesian model would add a prior distribution on d_{Ak} (as is the case for the fixed effects Bayesian model), and also on σ.

6.4 Model Validation

6.4.1 Traditional Meta-Analysis

The validity of the results based on the above procedures, both traditional meta-analysis and NMA, is dependent on a number of crucial assumptions.

A key assumption in all meta-analytic models, including traditional meta-analysis (a well as NMA, for that matter), is the requirement that treatment effects are relatively homogenous across trials involving the same pair of treatments. For various reasons, treatment effects are not expected to be completely homogenous across trials. The question then becomes whether the treatment effects are sufficiently homogenous to engender valid inferences about treatment effects. In the following, a few of the several ways of evaluating homogeneity are highlighted (Cappelleri et al., 2010).

One common approach is the chi-square test of homogeneity, defined earlier (Fleiss et al., 2003). However, in most typical applications of meta-analysis in biomedical fields, the number and size of studies are too limited and the chi-square test is thus underpowered. A nonsignificant result may not rule out heterogeneity (Ioannidis, 2008). Accordingly, as an arbitrary rule of thumb, *p*-values at or below 0.1 are usually taken to mean that the differences in effect sizes across studies should not be ignored.

As a function of the chi-square statistic of homogeneity (Q), the I^2 statistic is a descriptive measure that quantifies the extent of total variation in treatment effects across studies that is due to heterogeneity alone rather than chance (Higgins et al., 2002, 2003). In some respect, this test statistic is a kind of signal-to-noise ratio, and is computed as

$$I^2 = \left(\frac{Q - df}{Q} \right) 100\%$$

where *df* refers to degrees of freedom, which equals the number of studies minus 1. Values on I^2 can range from 0% to 100%. The proposed benchmark values on I^2 of 25%, 50%, and 75% might be considered low, medium, and high, respectively. The I^2 statistic is viewed as a measure of inconsistency across the findings of the studies.

A caveat about the I^2 is that it tends to have large uncertainty when there is a limited number of studies. The 95% confidence intervals of I^2 can be readily estimated and are useful to convey the extent of uncertainty (Ioannidis et al., 2007). One may also formally examine in sensitivity analyses the impact of specific studies on the extent of estimated between-study heterogeneity (Patsopoulos et al., 2008). As stated earlier, in the random effects approach, the heterogeneity is formally incorporated in the model. In Section 6.5, an alternative approach in the form of meta-regression, which involves incorporating study-level covariates, will be discussed.

6.4.2 NMA

NMA, whether frequentist or Bayesian, or whether fixed or random effects, has three principal assumptions (Donegan et al., 2013; Jansen et al., 2011, 2014; Lu and Ades, 2006; Salanti et al., 2014; Snedecor et al., 2014; Song et al., 2003). The assumption of *homogeneity* occurs, as in traditional meta-analysis, when the true treatment effects in a direct comparison of two treatments across studies are the same. Its negation or heterogeneity, also referred to as treatment-by-study interaction (effect modification), relates to the extent to which the true treatment effect varies according to populations/patient characteristics, treatment characteristics (such as dose or duration), or study characteristics.

Transitivity (or *similarity* or *exchangeability*), the second assumption, requires that the distribution of patient and study characteristics that are modifiers of the treatment effect be sufficiently similar in different sets of randomized controlled studies that go into an indirect comparison. If so, the relative effect estimated by trials of A versus C is generalizable to patients in trials of B versus C (and vice versa). In addition to clinical similarity, methodological similarity (e.g., quality, definition of outcomes) is required for valid estimates. If there is an imbalance in the distribution of effects modifiers (treatment-by-covariate interactions) between trials, then the estimates become biased. The assumption of transitivity implies that the same effect size would be obtained if the A versus B comparison was performed under the conditions of the B versus C comparative trial and vice versa—that is, the relative efficacy of treatments is the same in all trials included in the indirect comparison. This would be the case if one could demonstrate both methodological similarities as well as the comparability of patient populations across studies. Therefore, in an attempt to assess similarity, which is not directly testable, one may need to rely on qualitative as well as quantitative evaluations of comparability of relevant covariates and other trial attributes, at least for those that are observed, across studies (Alemayehu, 2011).

Consistency, the third assumption, refers to an agreement between direct and indirect evidence for a given pair of treatments. A lack of consistency is related to lack of transitivity, as inconsistency stems from a lack of transitivity. When the network involves both direct evidence and indirect evidence, the consistency of the data from the two sources should be verified. Alternative approaches have been proposed to assess consistency (Dias et al., 2010, 2013; Katsanos, 2014; Lu and Ades, 2006; Lumley, 2002).

The simplest method for testing the consistency of evidence, often referred to as the Bucher method, involves a comparison of a direct estimate of (say) treatment C versus treatment B and its corresponding indirect estimate from the AB and AC direct evidence for single loops of evidence (Dias et al., 2013; Katsanos, 2014). A Z test on the null hypothesis of no inconsistency can be calculated, involving the difference between these two estimates in the numerator and its standard error in the denominator. Although this particular illustration of the method can only be applied to three independent sources of data (loop ABC), the method can be generalized to other single loops of evidence (e.g., loop ABCD). These methods, though, are based on two-arm trials. Three-arm trials cannot be included because they are internally consistent and will therefore reduce the chance of inconsistency being detected.

Trials with two or more treatment arms can be addressed using methods for general networks (Dias et al., 2013; Katsanos, 2014). For more complicated networks, where statistically independent tests cannot be constructed for the different loops of evidence [e.g., four three-treatment loops (ACD, BCD, ABD, ABC) is not statistically independent from three four-treatment loops (ABCD, ACDB, CABD) in the same overall network], methods for general networks are available (Dias et al., 2013; Katsanos, 2014). One of them, called the unrelated means effects (UME) approach, uses a Bayesian framework and involves comparing the overall network model that incorporates the standard assumption of consistency (the standard consistency model) with an UME model (also called the inconsistency model), in which each of the pairwise treatment contrasts for which evidence is available represents a separate, unrelated, basic parameter to be estimated. No consistency is assumed in the UME model, which can use fixed or random effects. Results (estimates, standard deviation, credible interval) from the two types of models—one that assumes consistency and one that does not—can be compared. Comparison between the deviance information criterion (DIC) statistics of the consistency and UME models provides an omnibus test of consistency. Lower values of DIC provide a better fit.

Other general methods exist for detecting inconsistency (Dias et al., 2013; Katsanos, 2014). One approach is called node splitting, which is also implemented in a Bayesian framework. Node splitting allows the user to split the information contributing to estimates of a parameter (node), say the overall estimates between treatment A and treatment B, into two distinct components: the direct, based on all the AB data (which may come from AB, ABC, DAB trials); and the indirect, based on all the remaining evidence. Node splitting can also provide the opportunity to generate intuitive graphics based on direct, indirect, and combined evidence.

6.4.3 Publication Bias

Publication bias is induced when the decision to publish is influenced by whether the study result of an experimental treatment is "positive" (i.e., favorable and statistically significant) or "negative" (i.e., statistically unfavorable or not statistically significant) (Cappelleri et al., 2010). Theoretical and empirical evidence suggests that, if the treatment effects of the studies included in the meta-analysis are found to be related to the sample size or the variance of the treatment effects, then this association on some occasions may suggest publication bias (Sutton, 2009), as small negative studies may be more likely to be unpublished as compared with larger ones.

Several tests have been suggested in the literature to detect the possibility of publication bias, and some of them, such as the "funnel plot," in which a measure of precision (e.g., sample size or the reciprocal of the variance) is plotted against the treatment effect, are popular and commonly applied (Sutton, 2009). Nonparametric (Begg and Mazumdar, 1994) and parametric methods (Egger et al., 1997) to formally test for such funnel asymmetry have become accepted.

When selection bias is present, the plot becomes asymmetrical, and the overall effect of the meta-analysis is biased.

Empirical research, though, indicates that the shape of a funnel plot may be also determined by the arbitrary choice of the method used to construct the plot (i.e., by the definition of precision and the effect measure) (Tang and Liu, 2000). Asymmetry in a funnel plot may also stem from the heterogeneity of the treatment effect across studies, as well as from chance and subjectivity (Terrin et al., 2003). Therefore, judicious care and discernment should be given when interpreting a funnel plot.

In mitigating publication bias, research registries and other reasonable means (e.g., following up on published abstracts) can be used to track down unpublished studies. One reliable way to circumvent publication bias is probably via the use of prospective trial registries, because new trials are registered before they are started, and the decision to register a trial cannot be affected by the results of the study. Such registries can also address publication lag, the time lag that occurs in the publication of negative findings (relative to the publication of positive findings) after the completion of the trial follow-up.

A volume on publication bias in meta-analysis provides comprehensive coverage that includes the following: different types of publication bias, mechanisms that may induce them, empirical evidence for their existence, statistical methods to address them, and ways in which they can be avoided (Rothstein et al., 2005). This edited volume features worked examples with common datasets, and compares available software used for analyzing and reducing publication bias.

6.5 Meta-Regression

Meta-regression models include study-level covariates, and provide a way to adjust for or evaluate the extent to which these covariates account for the heterogeneity of treatment effects (Baker et al., 2009; Berlin et al., 1994; Sutton and Higgins, 2008). In a meta-regression of summary data, the unit of analysis is the individual study in the meta-analysis. Aggregate quantities reported in the studies, such as the mean dosage, mean age, and mean duration of disease, could be extracted and used as independent variables. The treatment effect (e.g., the risk difference) can be the dependent variable. Meta-regressions have been applied to several clinical areas (Barza et al., 1996; Cooper et al., 2009b; Donegan et al., 2012; Ioannidis et al., 1995).

Meta-regression has been described as the merging of meta-analytic techniques with linear regression principles (predicting treatment effects using covariates) (Sutton and Higgins, 2008). Meta-regression explores whether an association exists between variables and a treatment effect, along with the direction of that association. Meta-regression is a more sophisticated method than subgroup analysis for exploring heterogeneity, and has the potential advantage of efficiently allowing for the evaluation of one or more covariates simultaneously.

Taking the natural logarithm of a treatment effect measured on a multiplicative scale, such as the risk ratio and odds ratio, is suggested to help ensure that the dependent variable becomes normally distributed in order to make valid statistical inferences. It is more desirable to weigh each study in the meta-regression proportional to its precision (e.g., the inverse of the variance of a study's treatment effect), rather than weighing the studies equally.

Unfortunately, the number of studies in a traditional or NMA is often limited and, in such cases, adjustment by incorporating study-level covariates with meta-regression models may sometimes be questionable (Lu and Ades, 2006; Thompson et al., 1997). In addition, an aggregate-level covariate adjustment might produce ecological bias, limiting the interpretation of estimated results for subgroups (Berlin et al., 2002; Thompson and Higgins, 2002; Higgins and Thompson, 2004); what is related at the group level (e.g., the relationship between average age and treatment effect) may or may not be related at the individual level (e.g., the relationship between individual ages and treatment effect).

In contrast, individual patient-level meta-analyses (be them traditional or network) usually have sufficient power to validly and reliably estimate effects in meta-regression models, thereby reducing heterogeneity (for traditional meta-analysis or NMA) and inconsistency (for NMA), as well as providing the opportunity to explore differences in effects among subgroups (Donegan et al., 2012; Lambert et al., 2002; Stewart et al., 2015). However, obtaining patient-level data for all RCTs for a traditional meta-analysis, and especially an NMA, may be considered unrealistic. As an alternative, one could use patient-level data when available and aggregate level data for studies in the network for which such data is not available, thereby improving parameter estimation over aggregate-data-only models (Jansen, 2012; Pigott et al., 2012).

6.6 Best Practices for the Conduct and Reporting of Meta-Analysis

Several best practices are available for the conduct and reporting of meta-analysis (Baker et al., 2016). In this section, a few of the more prominent ones are highlighted.

6.6.1 Cochrane Collaboration's Risk of Bias Assessment: A Tool for Evaluating the Quality of Individual Randomized Trials

The Cochrane Collaboration's tool for assessing bias in randomized trials gives a useful classification of the different forms of bias: selection, performance, detection, attrition, and reporting bias with corresponding domains that are assessed by the tool for individual studies within a systematic review (Higgins et al., 2011b). The tool was developed to adhere to seven principles: (1) does not use quality scales; (2) only focuses on internal validity; (3) assesses the risk of bias in trial results, not the quality of reporting or methodological problems; (4) requires reviewer judgment; (5) assesses domains based on a combination of theoretical and empirical consideration; (6) focuses on risk of bias in the data as represented in the review, rather than other sources; and (7) reports outcome-specific evaluations of risk of bias. For each of the risk of bias domains, the assessment is done in two parts. The first describes the relevant trial characteristics on which the risk of bias is based. The second assigns a judgment of "low risk," "high risk," or "unclear risk" of bias.

As an extension to the Cochrane Collaboration's tool, another instrument was advanced to assess the quality of a meta-analysis that provides greater granularity and response categories, with an emphasis on the evaluation of statistical and interpretational issues (Higgins et al., 2013). This tool consists of 43 items divided into four key categories: data sources, analysis of individual studies by the meta-analyst, general meta-analysis, and reporting and interpretation, with a summary judgment question appearing at the end of each category. Accompanying the tool is detailed guidance for completing the assessment form.

6.6.2 PRISMA Suite, MOOSE, and MAER-Net: Tools for Reporting

In 1999, to address suboptimal reporting of meta-analyses, an international group published the QUality Of Reporting Of Meta-analyses (QUOROM) statement (Moher et al., 1999). Many journal editors and authors, including those involved in the Cochrane Collaboration, then pursued compliance with the QUOROM checklist to ensure that authors reported transparently what they did (methods) and found (results). In 2009, the QUOROM statement was updated to address several conceptual, methodological, and practical advances, and was renamed the Preferred Reporting Items of Systematic reviews and Meta-Analyses (PRISMA) statement (Moher et al., 2009).

The PRISMA statement consists of a 27-item checklist and a four-phase flow diagram. The checklist includes items deemed essential for transparent reporting of a systematic review or meta-analysis across seven sections: title, abstract, introduction, methods, results, discussion, and funding. In addition, the explanation and elaboration document is intended to enhance the use and understanding of the PRISMA statement, providing meaning and rationale for each checklist item through examples and explanations (Liberati et al., 2009).

Although PRISMA focuses on randomized trials, it can be used as a basis for reporting other types of research designs that evaluate interventions. While the PRISMA statement assesses the quality of the *reporting* of a published systematic review or meta-analysis, it is not an instrument to gauge the quality of the underlying systematic review *per se*.

Systematic reviews should build on a detailed, well-described, protocol. To date, few protocols are published or are mentioned in published systematic reviews. The Preferred Reporting Items for Systematic Reviews and Meta-Analyses Protocols (PRISMA-P) 2015 statement consists of a 15-item checklist to help develop and report a systematic review protocol (Moher et al., 2015) Again, an explanation and elaboration document provides an understanding of the necessity of each item and a model example (Shamseer et al., 2015).

There are two further specialized extensions to the PRISMA guidelines. PRISMA-Abstracts provides a checklist to give authors a structured framework for condensing a systematic review or meta-analysis into the essentials required for good reporting in journal and conference abstracts (Beller et al., 2013). PRISMA-Equity takes the PRISMA 27-item checklist and extends the item descriptions to give reporting guidelines for systematic reviews of effects on inequities in health outcomes and health care use across socioeconomic groups and other characteristics, with an aim to improve global health equity (Welch et al., 2012).

As an extension of the PRISMA for traditional pairwise treatment comparison (only two treatments given), a modified 32-item PRISMA extension checklist was developed to address what the group considered to be immediately relevant to the reporting of network meta-analyses (Hutton et al., 2015). Current PRISMA items were also clarified. This document presents the extension and provides examples of good reporting, as well as elaborations regarding the rationale for new checklist items and the modification of previously existing items from the PRISMA statement. It also highlights educational information related to key considerations in the practice of NMA.

The Meta-Analysis of Observational Studies in Epidemiology (MOOSE) group developed a checklist for preferred reporting of meta-analyses of nonrandomized evidence (Stroup et al., 2000). The MOOSE checklist is organized into background, search strategy, methods, results, discussion, and conclusions.

Meta-regression has expanded to the field of economics. The Meta-Analysis of Economics Research-Network (MAER-Net) has created reporting guidelines that center on research questions and effect size; research literature searching, compiling, and coding; and modeling issues (Stanley et al., 2013).

6.6.3 AMSTAR and GRADE: Tools to Appraise the Quality of Systematic Reviews

The most frequently recommended tool for assessing the methodological quality of a meta-analytic systematic review is the Assessment of Multiple Systematic Reviews (AMSTAR) (Shea et al., 2007). A final list of 11 components resulted and covers whether there was (1) a prior design, (2) any duplication of study selection and data extraction, (3) a comprehensive literature search, (4) use of status of publication (i.e., grey literature) as an inclusion criteria, (5) a list of studies included and excluded, (6) a set of characteristics of included studies, (7) assessment and documentation of the scientific quality of the included studies, (8) sufficient scientific quality of the individual studies to support the overall conclusions, (9) appropriate meta-analytic methods, (10) information on likelihood of publication bias, and (11) disclosure of conflicts of interest. Each of these 11 components is answered with a response of either "yes," "no," "can't answer," or "not applicable."

The AMSTAR tool can be used to assess the quality of included systematic reviews of both RCTs and nonRCTs. It gives, however, little or no consideration to the strength of the evidence supporting the conclusions of the systematic reviews.

For this purpose, the Grades of Recommendation, Assessment, Development and Evaluation (GRADE) provides a framework for rating quality of evidence and grading strength of recommendations in health care (Balshem et al., 2011). The GRADE approach is to rate the quality of a body of evidence for each main outcome of interest, not that of individual studies, and, within the context of a systematic review, GRADE reflects how confident we are that an effect estimate is close to the truth. GRADE rates the quality of evidence for each outcome across studies on a four-level scale. Randomized trials start at the high-quality level, while observational studies start at the low-quality level. Each outcome in each trial is then assessed against five reasons to downgrade and three reasons to upgrade. A final rating of quality for each outcome is given across studies.

6.6.4 Tools to Appraise the Quality of Systematic Reviews

The National Institute for Health Care Excellence Decision Support Unit commissioned seven technical support documents (TSD) on evidence synthesis for decision-making, including a reviewers' checklist (Ades et al., 2013). The checklist is intended for use with traditional pairwise meta-analysis, indirect comparisons, and NMA, without distinction, as the TSD series views NMA as an extension of pairwise meta-analysis with the assumptions of trial similarity and consistency in NMA being just properties of the identical/exchangeability requirement across all studies contributing to the relative treatment effects between a pair of treatments. The checklist consists of four sections: (1) definition of the decision problem, with an emphasis on effect modifiers; (2) methods of analysis and presentation of results; (3) issues specific to network synthesis; and (4) embedding the synthesis in a probabilistic cost- effectiveness analysis.

Similar to other systematic review guidelines, it does not generate a quality rating; it is not prescriptive and requires the reviewer to make a series of judgments. But, unlike other guidances, it provides a framework for open discussion and assumes reviewers can ask for clarification, alternate and sensitivity analyses, details of search algorithms, computer code, and so forth.

The International Society for Pharmacoeconomics and Outcomes Research (ISPOR) has sponsored two publications on good clinical research practices on NMA (Jansen et al., 2011; Hoaglin et al., 2011). More recently, ISPOR, the Academy of Managed Care Pharmacy (AMCP), and the National Pharmaceutical Council (NPC) collaborated to publish an article titled "Indirect treatment comparison/NMA study questionnaire to assess relevance and credibility to inform health care decision-making: An ISPOR-AMCP-NPC Good Practice Task Force Report" (Jansen et al., 2014). Development of the questionnaire was characterized by two principal concepts: relevance and credibility.

Another group of researchers proposed an approach to evaluate the quality of evidence from an NMA based on the GRADE working group methods for pairwise meta-analysis (Salanti et al., 2014). In extending GRADE to NMA, these researchers take a two-staged approach. First, they consider each GRADE domain (reason to downgrade: study limitations [risk of bias], indirectness, inconsistency, imprecision, and publication bias) separately for the available direct comparisons in the network. Second, they combine the domain-specific judgments for each pairwise network comparison to obtain the overall quality of evidence.

In the same timeframe, a GRADE working group published a four-step approach for rating the quality of treatment effect estimates from NMA (Puhan et al., 2014). The first step is to present both direct and indirect treatment estimates for each comparison in the network. The second step is to rate the quality of each of these direct and indirect comparisons. In the third step, the mixed treatment comparison estimates (both direct and indirect effects) are presented for the network.

The fourth and final step is to rate the quality of the NMA effect estimates.

6.7 Random Effects Meta-Analysis: A Simulated Example Using SAS®

Consider a simulated example based on 10 hypothetical studies involving the same pair of treatments in a traditional pairwise meta-analysis framework. The outcome variable is a patient-reported outcome measure, with the effect size expressed as the difference between mean scores. Using the previous notation, a random effects model can be expressed as

$$\hat{\theta}_l = \theta + p_i + e_i \tag{6.1}$$

where $\hat{\theta}_i$ is the estimated effect size for study i (i=1, 2, ..., 9, 10); θ is the common treatment effect; p_i is the residual heterogeneity in treatment effect for study i, which also can be described as the between-study variability, assumed to be from a normal distribution with mean 0 and variance σ_p^2 (i.e., $p_i \sim N(0, \sigma_p^2)$); and e_i is the within-study measurement error for study i, assumed to be from a normal distribution with mean 0 and known and distinct variance σ_{ei}^2 for every study (i.e., $e_i \sim N(0, \sigma_{ei}^2)$).

Figure 6.1 presents the SAS® code to generate simulated data. The term p_i in Equation 6.1 is generated using the code line *pi= rannor(1)*sqrt(variance_p)*, and will produce a value taken from the same normal distribution with a variance of 0.1 for all studies. In contrast, the other term e_i will be different for every study not only by value, but those values will be taken from different distributions with predefined variances for every study. This measurement

```
/*
The rannor (arbitrary seed value) function returns a (pseudo) random number from a
(standard) normal distribution with mean 0 and standard deviation 1.
*/
options nofmterr nocenter pagesize=2000 linesize=256;
Data _MetaAnalysis_;
NumberOfStudies = 10;
a = 2; /*value for the overall mean*/
variance_p = 0.1; /*value for the between-studies variance*/
Array variance_e{1:10} (0.02 0.03 0.04 0.015 0.035 0.042 0.022 0.032 0.035 0.04);
/*values for the within-study variance*/
Do study_id=1 To NumberOfStudies;
pi= rannor(1)*sqrt(variance_p) ;
ei= rannor(3)*sqrt(variance_e[study_id]);
y = a + pi + ei ;
output;
End;
Keep y study_id;
Run;
```

FIGURE 6.1
SAS code to generate simulated data.

error, e_i, is represented by the code line $ei = rannor(3) *sqrt(variance\_e[study\_id])$. The simulated data is then given by the code line $y = a + pi + ei$, with y and a corresponding to $\hat{\theta}$ *and* θ, respectively, in Equation 6.1.

Figure 6.2 gives the implementation of a mixed effects model using the SAS MIXED procedure. When the MIXED procedure is used to perform meta-analysis, the MODEL, RANDOM, REPEATED, and PARMS statements should be specified in a particular way. The MODEL statement describes only the fixed effect (θ) in Equation 6.1, while the statement "RANDOM INTERCEPT/Subject=study_id;" corresponds to "p_i." In the MODEL statement, the intercept (as a fixed effect) does not need to be included, since it is included by default. However, in the RANDOM statement, all random effects including the intercept should be specified.

The statement "REPEATED/Group=study_id;" corresponds to "e_i" in Equation 6.1 and describes within-study measurement error. The "Group" option in the "REPEATED" statement instructs the MIXED procedure to estimate distinct variance-covariance structure for every group defined by variable "study_id." As a result, for every group, a new set of variance-covariance parameters will be estimated (although all of them will be of the same structure). It should be noted that we did not add the type of the variance-covariance structure in the "REPEATED" statement; therefore, the default TYPE=VC (only variance components included) will be used.

As there is only one observation for a group (which is a study in our example), the variance components' variance-covariance structure will be reduced to be just an individual variance of the term "e_i." Note that in the "RANDOM" statement, we need to use the option "Subject=study_id" to model the value of the single between-studies variance corresponding to the term "p_i." As is the case of the "REPEATED" statement, the default TYPE=VC is also used in the "RANDOM" statement by default, which also leads to the simplest variance-covariance structure represented by just one variance.

```
Proc MIXED DATA=_MetaAnalysis_ ;

CLASS study_id;

MODEL y=  / cl solution;

RANDOM INTERCEPT / Subject=study_id;

REPEATED / Group=study_id;

PARMS

(0.5)

(0.02) (0.03) (0.04) (0.015) (0.035) (0.042) (0.022) (0.032) (0.035) (0.04)

/eqcons=2 to 11;

RUN;
```

FIGURE 6.2
SAS code to implement the random effects meta-analysis model.

```
             Covariance Parameter Estimates

Cov Parm       Subject      Group            Estimate

Intercept      study_id                      0.08468
Residual                    study_id 1       0.02000
Residual                    study_id 2       0.03000
Residual                    study_id 3       0.04000
Residual                    study_id 4       0.01500
Residual                    study_id 5       0.03500
Residual                    study_id 6       0.04200
Residual                    study_id 7       0.02200
Residual                    study_id 8       0.03200
Residual                    study_id 9       0.03500
Residual                    study_id 10      0.04000



                    Solution for Fixed Effects

                    Standard
Effect    Estimate   Error   DF   t Value   Pr > |t|  Alpha   Lower    Upper

Intercept   2.0300   0.1073    9    18.92    <.0001    0.05   1.7874   2.2727
```

FIGURE 6.3
Covariance parameter and fixed effect estimates.

As there is only one observation per study, the variance corresponding to the term "e_i," taken as a known value, cannot be estimated. The default purpose of the PARMS statement is to define initial values for the variance-covariance structures parameters specified by the "RANDOM" and "REPEATED" statements.

Figure 6.3 shows the results of the simulation exercise. For example, the modeled variance for the random intercepts (i.e., between-studies variance) is 0.08468, which is the estimate of σ_p^2, and compares well with the simulated value of 0.1. Note that the within-study variances are kept the same without any changes as they were defined by PARMS statement; these are the pre-defined estimates of σ_{ei}^2 for the 10 studies ($i = 1, \ldots, 10$). The modeled overall mean is 2.0300, which corresponds well to the simulated value of 2.

While SAS was the software chosen for implementation here, other software could have been selected instead. Advances in software programs dedicated to meta-analysis have made the statistical computing of pooled data more accessible and powerful for the analysis of real data (Bax et al., 2007). For example, like SAS, the software R can be applied not only to implement simulations, but also to analyze real data (Chen and Peace, 2015).

6.8 Discussion

Meta-analysis may be defined as the statistical analysis of data from multiple studies. A meta-analysis typically identifies data systematically, summarizes results, and evaluates quantitatively sources of heterogeneity and bias. This chapter highlighted the benefits and limitations of meta-analysis. Statistical frameworks, both frequentist and Bayesian, were described for pooling results using traditional meta-analysis and NMA. Cumulative meta-analysis was noted as a way to combine information from studies included in a sequential manner according to a certain ordered metric, such as publication year.

In the absence of individual patient data, the analysis of aggregate data is an important component in generating the best-available evidence to advance medical science and promote public health. However, as discussed in earlier sections, the approach is built on important assumptions that require careful evaluation. While there are statistical techniques galore to analyze aggregate data, it should be noted that the approaches are only as good as the validity of the assumptions upon which they are dependent. In some instances, the assumptions may not be testable, as is the case with the issue of similarity in NMA. Nonetheless, the robustness of results to departure from any assumptions should be evaluated, and sensitivity analyses performed, before the results are used in decision-making with respect to efficacy, safety, cost and quality of life, which impacts cost-effectiveness evaluations.

In recognition of the issues associated with the analysis of aggregate data, guidelines for best practices are continually published and refined as the literature evolves. Integral in most guidelines is the need for careful planning and adherence to good research practice, including formulation of pre-specified hypotheses, especially when the goal is to generate confirmatory evidence, protocols, and statistical analysis plans for data collection and analysis, and report results with fair balance.

References

Ades, A.E., Caldwell, D.M., Reken, S. et al. 2013. Evidence synthesis for decision making 7: A reviewer's checklist. *Med Decis Making* 33:679–691.

Afilalo, J., Duque, G., Steel, R. et al. 2008. Statins for secondary prevention in elderly patients: A hierarchical Bayesian meta-analysis. *J Am Coll Cardiol* 51:37–45.

Alemayehu, D. 2011. Assessing exchangeability in indirect and mixed treatment comparisons. *Comp Eff Res* 1:51–55.

Antman, E.M., Lau, J., Kupelnick, B. et al. 1992. A comparison of results of meta-analyses of randomized control trials and recommendations of clinical experts. Treatments for myocardial infarction. *JAMA* 268:240–248.

Baker, W.L., Bennetts, M., Coleman, C.I. et al. 2016. Appraising evidence. In *Umbrella Reviews: Evidence Synthesis with Overviews of Reviews and Meta-Epidemiologic Studies*, ed. G. Biondi-Zoccai, pp. 115–136. Switzerland: Springer International Publishing.

Baker, W.L., White, C.M., Cappelleri, J.C. et al. 2009. A clinical perspective on the use of meta-regression in systematic reviews: Caveats and cautions. *Int J Clin Pract* 63:1426–1434.

Balshem, H., Helfand, M., Schunemann, H.J. et al. 2011. GRADE guidelines: 3. Rating the quality of evidence. *J Clin Epidemiol* 64:401–406.

Barza, M., Ioannidis, J.P.A., Cappelleri, J.C. et al. 1996. Single or multiple daily doses of aminoglycosides: A meta-analysis. *BMJ* 312:338–345.

Bax, L., Yu, L.-M., Ikeda, N. et al. 2007. A systematic comparison of software dedicated to meta-analysis causal studies. *BMC Med Res Methods* 7:40. Open Access.

Begg, C.B. and M. Mazumdar. 1994. Operating characteristics of a rank correlation test for publication bias. *Biometrics* 50:1088–1101.

Beller, E.M., Glasziou, P.P., Altman, D.G. et al. April 2013. PRISMA for abstracts: Reporting systematic reviews in journal and conference abstracts. *PLoS Med* 10(4): e1001419. http://journals.plos.org/plosmedicine/article?id=10.1371/journal.pmed.1001419 (accessed May 29, 2017).

Berkey, C.S., Mosteller, F., Lau, J. et al. 1996. Uncertainty of the time of first significance in random effects cumulative metaanalysis. *Control Clin Trials* 17:357–371.

Berlin, J.A. and E.M. Antman. June 4, 1994. Advantages and limitations of metaanalytic regressions of clinical trials data. *Online J Curr Clin Trials*, Doc No 134.

Berlin, J.A., Santanna, J., Schmid, C.H. et al. 2002. Individual patient- versus group-level data meta-regressions for the investigation of treatment effects modifiers: Ecological bias rears its ugly head. *Stat Med* 21:3713–3787.

Borenstein, M., Hedges, L.V., Higgins, J.P.T. et al. 2009. *Introduction to Meta-Analysis.* Chichester: John Wiley & Sons.

Bucher, H.C., Guyatt, G.H., Griffith, L.E. et al. 1997. The results of direct and indirect treatment comparisons in meta-analysis of randomised controlled trials. *J Clin Epidemiol* 50:683–691.

Cappelleri, J.C., Ioannidis, J.P.A. and J. Lau. 2010. Meta-analysis of therapeutic trials. In: *Encyclopedia of Biopharmaceutical Statistics: 3rd Edition, Revised and Expanded*, ed. S.-C. Chow, pp. 768–779. New York, NY: Informa Healthcare.

Chen, D.-G. and K.E. Peace. 2015. *Applied Meta-Analysis with R.* Boca Raton, FL: Chapman & Hall/CRC Press.

Cipriani, A., Furukawa, T.A., Salanti, G. et al. 2009. Comparative efficacy and acceptability for 12 new-generation antidepressants: A multiple-treatments meta-analysis. *Lancet* 373:746–758.

Clarke, M., Brice, A. and I. Chalmers. July 2014. Accumulating research: A systematic account of how cumulative meta-analyses would have provided knowledge, improved health, reduced harm and saved resources. *PLOS ONE* 9(7):e102670. http://journals.plos.org/plosone/article/asset?id=10.1371%2Fjournal.pone.0102670.PDF (accessed May 30, 2017). doi:10.1371/journal.pone.0102670

Cooper, H., Hedges, L.V. and J.C. Valentine, eds. 2009a. *The Handbook of Research Synthesis and Meta-Analysis*, 2nd edition. New York, NY: Russell Sage Foundation.

Cooper, N.J., Sutton, A.J., Morris, D. et al. 2009b. Addressing between-study heterogeneity and inconsistency in mixed treatment comparisons: Application to stroke prevention treatments in individuals with non-rheumatic atrial fibrillation. *Stat Med* 28:1861–1881.

DerSimonian, R. and N. Laird. 1986. Meta-analysis in clinical trials. *Control Clin Trials* 7:177–188.

Dias, S., Welton, N.J., Caldwell, D.M. et al. 2010. Checking consistency in mixed treatment comparison meta-analysis. *Stat Med* 29:932–944.

Dias, S., Welton, N.J., Sutton, A.J. et al. 2013. Evidence synthesis for decision making 4: Inconsistency in networks of evidence based on randomized controlled trials. *Med Decis Making* 33:641–656.

Donegan, S., Williamson, P., D'Alessandra, U. et al. 2012. Assessing the consistency assumption by exploring treatment by covariate interactions in mixed treatment comparison meta-analysis: Individual patient-level covariates versus aggregate trial-level covariates. *Stat Med* 31:3840–3857.

Donegan, S., Williamson, P., D'Alessandra, U. et al. 2013. Assessing key assumptions of network meta-analysis. *Res Synth Meth* 4:291–323.

Donegan, S., Williamson, P., Gamble, C. et al. November 2010. Indirect comparisons: A review of reporting and methodological quality. *PLOS ONE* 5(11):e11054–e11054. http://journals.plos.org/plosone/article?id=10.1371/journal.pone.0011054 (accessed May 30, 2017).

Egger, M. and G.D. Smith. 1995. Misleading meta-analysis. *BMJ* 310:752–754.

Egger, M., Smith, G.D. and D.G. Altman, eds. 2001. *Systematic Reviews in Health Care: Meta-Analysis in Context*, 2nd edition. London: BMJ Books.

Egger, M., Smith, G.D., Schneider M. et al. 1997. Bias in meta-analysis detected by a simple, graphical test. *BMJ* 315:629–634.

Fleiss, J.L., Levin, B. and C.M. Paik. 2003. *Statistical Methods for Rates and Proportions*, 3rd edition. Hoboken, NJ: John Wiley & Sons.

Hardy, R.J. and S.G. Thompson. 1996. A likelihood approach to meta-analysis with random effects. *Stat Med* 15:619–629.

Higgins. J.P.T., Altman, D.G. and J.A.C. Sterne. 2011a. Assessing risk of bias in included studies. In *Cochrane Handbook for Systematic Reviews of Interventions*, eds. J.P.T. Higgins and S. Green, Chapter 8, version 5.1.0. London: The Cochrane Collaboration. http://handbook.cochrane.org/

Higgins, J.P.T., Lane, P.W., Anagnostelis, B. et al. 2013. A tool to assess the quality of a meta-analysis. *Res Synth Meth* 4:351–366.

Higgins, J.P.T. and S.G. Thompson. 2002. Quantifying heterogeneity in a meta-analysis. *Stat Med* 21:1539–1558.

Higgins J.P.T. and S.G. Thompson. 2004. Controlling the risk of spurious findings from meta-regression. *Stat Med* 23:1663–1682.

Higgins, J.P.T., Thompson, S.G., Deeks, J.J. et al. 2003. Measuring inconsistency in meta-analyses. *BMJ* 327:557–560.

Higgins, J. P. T., Whitehead, A. and M. Simmonds. 2011b. Sequential methods for random-effects meta-analysis. *Stat Med* 30(9), 903–921. http://doi.org/10.1002/sim.4088 (accessed May 30, 2017).

Hoaglin, D.C., Hawkins, N., Jansen, J.P. et al. 2011. Conducting indirect-treatment-comparison and network-meta-analysis studies: Report of the ISPOR Task Force on Indirect Treatment Comparisons Good Research Practices—Part 2. *Value Health* 14:429–437.

Hu, M., Cappelleri, J.C. and K.K.G. Lan. 2007. Applying the law of iterated logarithm to control type 1 error in cumulative meta-analysis of binary outcomes. *Clin Trials* 4:329–340.

Hutton, B., Salanti, G., Caldwell, D.M. et al. 2015. The PRISMA extension statement for reporting of systematic reviews incorporating network meta-analyses of health care interventions: Checklist and explanations. *Ann Intern Med* 162:777–784.

Ioannidis, J.P. 2008. Interpretation of tests of heterogeneity and bias in meta-analysis. *J Eval Clin Pract* 14:951–957.

Ioannidis, J.P.A., Cappelleri, J.C., Lau J. et al. 1995. Early or deferred zidovudine in HIV-infected patients without an AIDS-defining illness. A meta-analysis. *Ann Intern Med* 122:856–866.

Ioannidis, J.P., Patsopoulos, N.A. and E. Evangelou. 2007. Uncertainty in heterogeneity estimates in meta-analyses. *BMJ* 335:914–916.

Jansen, J.P. 2012. Network meta-analysis of individual and aggregate level data. *Res Synth Meth* 3:177–190.

Jansen, J.P., Fleurence, R., Devine, B. et al. 2011. Interpreting indirect treatment comparisons and network meta-analysis for health-care decision making: Report of the ISPOR Task Force on Indirect Treatment Comparisons Good Research Practices: Part 1. *Value Health* 14:417–428.

Jansen, J.P., Trikalinos, T., Cappelleri, J.C. et al. 2014. Indirect treatment comparison/network meta-analysis study questionnaire to assess relevance and credibility to inform health care decision making: An ISPOR-AMCP-NPC Good Practice Task Force Report. *Value Health* 17:157–173.

Katsanos, K. 2014. Appraising inconsistency between direct and indirect estimates. In *Network Meta-Analysis: Evidence Synthesis with Mixed Treatment Comparison*, ed. G. Biando-Zoccai, pp. 191–210. New York, NY: Nova Science Publishers.

Lan, K.K.G., Hu, M. and J.C. Cappelleri. 2003. Applying the law of iterated logarithm to cumulative meta-analysis of a continuous endpoint. *Stat Sinica* 13:1135–1145.

Lambert, P.C., Sutton, A.J., Abrams, K.R. et al. 2002. A comparison of summary patient-level covariates in meta-regression with individual patient data meta-analysis. *J Clin Epidemiol* 55:86–94.

Lau, J., Antman, E.M., Jimenez-Silva, J. et al. 1992. Cumulative meta-analysis of therapeutic risk for myocardial infarction. *N Eng J Med* 327:248–254.

Lau, J., Schmid, C.H. and T.C. Chalmers. 1995. Cumulative meta-analysis of clinical trials builds evidence for exemplary medical care. *J Clin Epidemiol* 48:45–57.

Liberati, A., Altman, D.G., Tetzlaff, J. et al. 2009. The PRISMA statement for reporting systematic reviews and meta-analyses of studies that evaluate health care interventions: Explanation and elaboration. *Ann Intern Med* 151:W-65–W-94.

Lu, G. and A.E. Ades. 2004. Combination of direct and indirect evidence in mixed treatment comparisons. *Stat Med* 23:3105–3124.

Lu, G. and A.E. Ades. 2006. Assessing evidence inconsistency in mixed treatment comparisons. *J Am Stat Assoc* 101:447–459.

Lumley, T. 2002. Network meta-analysis for indirect treatment comparisons. *Stat Med* 21:2313–2324.

Mayo-Wilson, E., Hutfless, S., Li, T. et al. 2015. Integrating multiple data sources (MUDS) for meta-analysis to improve patient-centered outcomes research: A protocol for a systematic review. *Syst Rev* 4:143. doi:10.1186/s13643-015-0134-z. http://link.springer.com/content/pdf/10.1186/s13643-015-0134-z.pdf (accessed May 30, 2017).

Moher, D., Cook, D.J., Eastwood, S. et al. 1999. Improving the quality of reports of meta-analyses of randomised controlled trials: The QUOROM statement. Quality of reporting of meta-analyses. *Lancet* 354:1896–1900.

Moher, D., Liberati, A., Tetzlaff, J. et al. July 2009. Preferred reporting items for systematic reviews and meta-analyses: The PRISMA statement. *PLoS Med* 6(7):e1000097. http://journals.plos.org/plosmedicine/article?id=10.1371/journal.pmed.1000097 (accessed May 30, 2017).

Moher, D., Shamseer, L., Clarke, M. et al. January 2015. Preferred reporting items for systematic review and meta-analysis protocols (PRISMA-P) 2015 statement. *Syst Rev* 4:1. http://systematicreviewsjournal.biomedcentral.com/articles/10.1186/2046-4053-4-1 (accessed May 30, 2017).

Patsopoulos, N.A., Evangelou, E. and J.P. Ioannidis. 2008. Sensitivity of between-study heterogeneity in meta-analysis: Proposed metrics and empirical evaluation. *Int J Epidemiol* 37:1148–1157.

Pigott, T.D., Williams, R. and J.R. Polanin. 2012. Combining individual participant and aggregated data in a meta-analysis with correlational studies. *Res Synth Methods* 3:257–268.

Puhan, M.A., Schünemann J.H., Murad, M.H. et al. 2014. A GRADE Working Group approach for rating the quality of treatment effect estimates from network meta-analysis. *BMJ* 349:g5630.

Quant, E.C., Jeste, S.S., Muni, R.H. et al. 2009. The benefits of steroids versus steroids plus antivirals for treatment of Bell's palsy: A meta-analysis. *BMJ* 339:b3354.

Rothstein, H.R., Sutton, A. J. and M. Borenstein, eds. 2005. *Publication Bias in Meta-Analysis: Prevention, Assessment and Adjustments*. Chichester: John Wiley & Sons.

Sacks, H.S., Berrier, J., Reitman, D. et al. 1987. Meta-analyses of randomized controlled trials. *N Engl J Med* 316:450–455.

Salanti, G., Del Giovane, C., Chaimani, A. et al. July 2014. Evaluating the quality of evidence from a network meta-Analysis. *PLOS ONE* 9(7):e99682. http://journals.plos.org/plosone/article?id=10.1371/journal.pone.0099682 (accessed May 30, 2017).

Schmid, J.E., Koch, G.G. and L.M. LaVange. 1991. An overview of statistical issues and methods of meta-analysis. *J Biopharm Stat* 1:103–120.

Senn, S., Gavini, F., Magrez, D. et al. 2013. Issues in performing a network meta-analysis. *Stat Methods Med Res* 22:169–189.

Shamseer, L., Moher, D., Clarke, M. et al. 2015. Preferred reporting items for systematic review and meta-analysis protocols (PRISMA-P) 2015: Elaboration and explanation. *BMJ* 349:g7647.

Shea, B.J., Grimshaw, J.M., Wells, G.A. et al. February 2007. Development of AMSTAR: A measurement tool to assess the methodological quality of systematic reviews. *BMC Med Res Method* 7:10. http://bmcmedresmethodol.biomedcentral.com/articles/10.1186/1471-2288-7-10 (accessed May 30, 2017).

Snedecor, S.J., Patel, D.A. and J.C. Cappelleri. 2014. From pairwise to network meta-analyses. In *Network Meta-Analysis: Evidence Synthesis with Mixed Treatment Comparison*, ed. G. Zoccai-Biando, pp. 21–41. New York, NY: Nova Science Publishers.

Song, F., Altman, D., Glenny, A. et al. 2003. Validity of indirect comparison for estimating efficacy of competing interventions: Empirical evidence from published meta-analyses. *BMJ* 326:472–476.

Stanley, T.D., Doucouliagos, H., Giles, M. et al. 2013. Meta-analysis of economics research reporting guidelines. *J Econ Surv* 27:390–394.

Stewart, L.A., Clarke, M., Rovers, M. et al. 2015. Preferred reporting items for a systematic review and meta-analysis of individual participant data:The PRISMA-IPD Statement. *JAMA* 313:1657–1665.

Stroup, D.F., Berlin, J.A., Morton, S.C. et al. 2000. Meta-analysis of observational studies in epidemiology. A proposal for reporting. *JAMA* 283:2008–2012.

Sutton, A.J. 2009. Publication bias. In *The Handbook of Research Synthesis*, eds. H. Cooper, L.V. Hedges and J.C. Valentine, pp. 435–452. New York, NY: Russell Sage Foundation.

Sutton, A.J. and K.R. Abrams. 2001. Bayesian methods in meta-analysis and evidence synthesis. *Stat Methods Med Res* 10:277–303.

Sutton, A.J. and J.P.T. Higgins. 2008. Recent developments in meta-analysis. *Stat Med* 27:625–650.

Tan, S., Xiao, X., Ma, H. et al. August 2015. Clopidogrel and aspirin versus aspirin alone for stroke prevention: A meta-analysis. *PLOS ONE* 10(8):e0135372. http://journals.plos.org/plosone/article?id=10.1371/journal.pone.0135372 (accessed May 30, 2017).

Tang, J.-L. and J.L.Y. Liu. 2000. Misleading funnel plot for detection of bias in meta-analysis. *J Clin Epidemiol* 53:477–484.

Terrin, N., Schmid, C.H., Lau, J. et al. 2003. Adjusting for publication bias in the presence of heterogeneity. *Stat Med* 22:2113–2126.

Thompson, S.G. and J.P. Higgins. 2002. How should meta-regression analyses be undertaken and interpreted? *Stat Med* 21:1559–1573.

Thompson, S.G. and S.J. Pocock. 1991. Can meta-analysis be trusted? *Lancet* 338:1127–1130.

Thompson, S.G., Smith, T.C. and S.J. Sharp. 1997. Investigating underlying risk as a source of heterogeneity in meta-analysis. *Stat Med* 16:2741–2758.

Uman, L.S. 2011. Systematic reviews and meta-analyses. *J Can Acad Child Adolesc Psychiatry* 20:57–59.

Welch, V., Petticrew, M., Tugwell, P. et al. October 2012. PRISMA-equity 2012 extension: Reporting guidelines for systematic reviews with a focus on health equity. *PLoS Med* 9(10):e1001333. http://journals.plos.org/plosmedicine/article?id=10.1371/journal.pmed.1001333 (accessed May 30, 2017).

Wells, G.A., Sultan, S.A., Chen, L. et al. 2009. *Indirect Evidence: Indirect Treatment Comparisons in Meta-Analysis*. Ottawa: Canadian Agency for Drugs and Technologies in Health.

Whiting, P.F., Wolff, R.F., Deshpande, S. et al. 2015. Cannabinoids for medical use: A systematic review and meta-analysis. *JAMA* 313:2456–2473.

7

Health Economics and Outcomes Research in Precision Medicine

**Demissie Alemayehu, Joseph C. Cappelleri,
Birol Emir, and Josephine Sollano**

CONTENTS

7.1 Introduction 151
7.2 Economic Evaluation in Precision Medicine 153
7.3 Statistical Considerations... 154
    7.3.1 Biomarkers in Personalized Medicine.................................... 155
    7.3.2 Statistical Issues in Genomic Analysis 155
    7.3.3 Subgroup Analysis 156
7.4 PROs in Precision Medicine 157
    7.4.1 Instrument Development and Validation 159
    7.4.2 Longitudinal Models... 161
    7.4.3 Item Response Theory (IRT).. 162
    7.4.4 Missing Data with PROs.. 163
    7.4.5 Interpretation... 165
7.5 Regulatory, Access, and Other Considerations in
    Precision Medicine 166
    7.5.1 HTA... 166
    7.5.2 Regulatory Issues.. 167
    7.5.3 Ethical Issues 168
    7.5.4 Clinical Practice... 168
7.6 Concluding Remarks... 169
Acknowledgments 170
References.. 170

7.1 Introduction

Personalized medicine involves getting the right therapy to the right patient at the right time. It is often used with various terms including precision medicine, stratified medicine, and pharmacogenomics. In this chapter, the two terms *precision medicine* and *personalized medicine* will be used interchangeably.

Precision medicine has been defined by the United States (US) National Academy of Sciences as "the use of genomic, epi-genomic, exposure and other data to define individual patterns of disease, potentially leading to better individual treatment." (National Research Council, 2011). In this respect, the overarching goal of personalized medicine is to find treatments that target cells, molecules, genes, and so forth, as opposed to the traditional approach of focusing on organs or body systems.

From the standpoint of improving public health, there is a general consensus about the value of targeting treatment at patients who are highly likely to benefit from it. A core belief behind the concept of personalized medicine is that it may expedite drug development and lead to greater efficacies in health care (Couch and Moft, 2012). However, the assessment of the corresponding economic impact appears to be relatively less straightforward. For example, the usual economic evaluation of drugs is normally based on data obtained from studies conducted on a broad spectrum of patient populations.

On the other hand, the focus in personalized medicine is on the identification of patient characteristics that discriminate responders from nonresponders. Since most studies are not prospectively designed to address such evaluations, retrospective analyses aimed at subgroup identification may lack statistical robustness. Nevertheless, global payer systems and reimbursement authorities are increasingly interested in understanding these aspects of personalized medicine so as to determine which patients are more likely to benefit from the new therapy. Current practice, therefore, is moving toward the inclusion of *a priori* defined subgroup analyses in order to make such determinations.

Use of biomarkers that predict response is an important way of gaining efficiency in personalized medicine, especially with molecular targeted therapies that tend to involve a high cost of delivery (Soria et al., 2011). Nonetheless, the development of genomic biomarkers involves surmounting special statistical challenges, including handling of high dimensionality and multiplicity issues, and assessment of the validity and clinical utility of the biomarker. Recent advances in modern analytic methods and new generation sequencing appear to address some of these issues, but more research is needed in this area (Matsui, 2013).

The role of patient-reported outcomes (PROs) and clinical outcome assessments (COAs) in personalized medicine is also a growing area of research (see, e.g., Alemayehu and Cappelleri, 2012). As a tool to quantify the response to treatment as perceived by the individual, PROs require a careful assessment of the validity and reliability of the instruments, as well as the associated statistical issues, including multiple comparisons, missing data handling, and the implementation of nonstandard modeling approaches (Hays et al., 2005).

Recent advances in personalized medicine have presented considerable challenges and opportunities for traditional health economics and outcomes research (HEOR), with far-reaching implications for reimbursement and access (Faulkner et al., 2012; O'Donnell, 2013). In this chapter, we provide a general overview of some of these challenges and opportunities. Section 7.2 addresses economic considerations in precision medicine. Section 7.3 summarizes analytical issues and approaches pertaining to genomic biomarker and subgroup analysis. Section 7.4 discusses the role of PROs in personalized medicine. Section 7.5 provides perspectives on regulatory, access, and other aspects of personalized medicine. Section 7.6 adds concluding remarks. Throughout, selected illustrative examples are highlighted from the recent literature.

7.2 Economic Evaluation in Precision Medicine

There are several factors that determine the successful integration of a new pharmacogenomics technology for personalized medicine into clinical practice. A critical initial step is the determination of its clinical utility and, hence, the relative benefits and risks of the technology (Holtzman and Watson, 1997). However, robust data on the clinical utility of genetic tests may not always be readily available.

In addition to having proven clinical utility, the new technology should also demonstrate economic benefits that are rigorously established using state-of-the-science analytical techniques and quality data (Deverka et al., 2010). As discussed in Chapter 5 of this monograph, the economic evaluation may be executed using alternative approaches, including cost-effectiveness analysis and cost-utility analysis (Drummond et al., 2005). While the application of these techniques in genomic medicine is generally similar to such other types of health care delivery, there are certain modifications and refinements that should be incorporated to take into account the unique challenges posed by personalized medicine. These include the activities pertaining to genome sequencing, companion diagnostic development and use, data handling and processing, and the need for genetic counseling (Fragoulakis et al., 2015).

With regards to cost, some of the drivers that should be factored into the economic evaluation include genetic sample collection and testing, treatment delivery, and education and training. Outcome measures in genomic evaluations may also be expressed in terms of personal utility, a situation in which the genomic information may not have clinical utility but may impact the person's general well-being (Foster et al., 2009). For example, depending on the results of a genomic test, the patients' perception of their own health may change, thereby potentially impacting their lifestyle. However, personal utility cannot be accurately measured for use in cost-effectiveness analysis.

One of the most widely used techniques—incremental cost-effectiveness analysis—involves computing the ratio of incremental cost of a new technology to the associated incremental effectiveness (e.g., quality-adjusted life-years).

The approach is known to have fundamental conceptual and practical limitations (O'Brien and Briggs, 2002). In the context of genomic medicine, the problem is compounded by some of the inherent issues of economic evaluation of pharmacogenomic interventions. For optimal decision-making, it is therefore essential to enhance the traditional cost-effectiveness analytical approaches, incorporating other relevant factors, including budget impact analysis and social preferences (Fragoulakis et al., 2015).

Another key component of personalized medicine is use of companion diagnostics, which involves testing before giving treatment. From the standpoint of health economics, this is generally viable if the cost of screening is not in excess of the savings obtained by tailoring treatment only to those who benefit from it, or to avoid treating those who may be harmed by it. The evidence necessary to evaluate the cost-benefit of such treatment modalities may not always be readily available at launch time of a new medicinal product, especially during negotiations with pricing and reimbursement agencies. For example, molecular diagnostics, which are generally established based on analytical performance data, may not have been adequately studied in clinical trials, as is required in most health technology assessment (HTA) reviews.

7.3 Statistical Considerations

The selection of an optimal treatment strategy at the individual patient level depends on a systematic evaluation of a patient's genomic and clinical characteristics in relation to the biological mechanisms of the disease under study. This requires a judicious choice of the statistical procedures to be used in data analysis. While traditional univariate techniques can serve as screening tools, more sophisticated multivariate methods may need to be applied to handle data complexity, including potential correlations among genes and other variables of interest. Given the high-dimensional nature of most genomic data, which are often associated with missing values, multicollinearity, multiplicity, and other computational challenges, traditional data analysis techniques may not always satisfactorily work.

Modern analytical approaches, including penalized regression and other machine learning tools, may be viable options in situations where conventional techniques fall short (Ma et al., 2015). A demonstration of the application of a linear regression model to develop an individual treatment rule based on data from a randomized controlled trial (RCT) may be found in DeRubeis et al. (2014). An example of the use of a penalized regression approach in an AIDS trial is given in Lu et al. (2013). For a detailed discussion on some of these techniques, see Chapter 4 of this monograph.

7.3.1 Biomarkers in Personalized Medicine

Biomarkers, also called molecular markers or signature molecules, serve as a primary tool to identify patients who may or may not benefit from a given treatment. There are different types of biomarkers that vary by functional use. They include *predisposition biomarkers,* used to assess the risk of developing a disease; *diagnostic biomarkers,* concerned with the identification of specific diseases; *prognostic biomarkers,* used to evaluate the course of a disease; and *predictive biomarkers,* employed to predict response to a drug, and, hence, to identify patients who may benefit from a given treatment.

Depending on the intended use, several approaches are followed to validate a biomarker. Analytical validation is performed to establish the reliability of the assay and the sensitivity and specificity of the measurements, often with reference to a gold standard (Chau et al., 2008; Zou et al., 2011). The value of a biomarker for evaluating disease prognosis or predicting response to treatment effect is assessed by establishing its clinical validity. For a prognostic biomarker, this involves establishing a strong correlation between the biomarker and a clinical endpoint. When data on a clinical endpoint are available from RCTs, the clinical validity of a predictive biomarker may be evaluated through an assessment of the treatment-by-biomarker interaction term in a suitable statistical model. The clinical utility of the biomarker is another attribute that needs to be established to ensure that its use in clinical practice has benefits. When the focus is to establish clinical utility for a newly developed biomarker, the trial design may involve randomizing patients either to a standard-of-care therapy or to a strategy in which a biomarker-based treatment assignment is used.

Biomarkers can also be incorporated in trial designs, either to enhance efficiency or to identify subgroups of interest. In an enrichment design, only patients who are predicted to be responders based on the biomarker are used to compare a new treatment option versus a control. Such a design may be employed when there is evidence showing that test-negative patients are not likely to respond to treatment and thus it may be unethical to enroll these patients in the trial. For a discussion of alternative study designs and analytical approaches, see, for example, Gosho et al. (2012), Matsui (2013), and Simon (2010).

7.3.2 Statistical Issues in Genomic Analysis

Alternative statistical techniques are available for use in genomic studies (Amaratunga et al., 2014). Commonly, a genome-wide association study is conducted to investigate potential associations between single-nucleotide polymorphisms and a disease of interest. A general strategy for screening concerns performing multiple tests based on gene expression measurements using common univariate statistical procedures, including a chi-squared test, Fisher's exact test, logistic regression, or a two-sample t-test. In this scenario, a major issue is one of controlling false positive rates. The false discovery rate (FDR) (Benjamini and Hochberg, 1995) has been proposed as an alternative to more conservative traditional methods for simultaneous inference.

Subsequent enhancements of the FDR include a modified version proposed by Storey and Tibshirani (2001), called the positive false discovery rate (pFDR). The q-value is used as a measure of significance in terms of the FDR, rather than the usual false positive rate associated with traditional p-values (Storey and Tobshirani, 2003). An alternative approach, introduced in Kostis et al. (2013), involves simulated randomization to control FDR. In certain applications, the selection of candidate genes may also be accomplished by first ranking them with respect to the magnitude of association or effect sizes, and then retaining a small set of the top ranking genes (Pepe et al., 2003).

A limitation of univariate approaches is that they do not take into account potential correlations among genes. Accordingly, hierarchical models that draw strength by incorporating information across comparable genes have been proposed (Speed, 2003). After a promising set of genes is identified in the preliminary screening, the diagnostic potential of the markers is then further assessed using alternative models, including traditional or machine learning tools (Breiman, 2001; Trevor et al., 2009), while incorporating additional phenotypic information. Finally, the predictive accuracy of the model should be determined based on suitable validation techniques. These may include internal and external validation, depending on whether the data used for validation are internal or external to the study population, respectively (Molinaro et al., 2005).

Another issue—often referred to as "misinformation chaos"—pertains to the fact that in most applications, the number of truly informative genes is small relative to the pool of genes under consideration. In these cases, even ensemble classifiers, such as random forests, may perform poorly. Amaratunga et al. (2008) proposed an enrichment approach based on a weighting strategy such that the genes that are most likely to separate the groups are assigned relatively more weights. This enrichment idea has also been extended to other techniques (Amaratunga et al., 2014).

7.3.3 Subgroup Analysis

There is a growing body of literature on classification and regression techniques used in subgroup identification. The traditional approach is based on an assessment of treatment-by-subgroup interaction using a suitable linear or generalized linear model. In the presence of multicollinearity or high dimensionality, nonstandard techniques, such as penalized regression, may be preferred to conventional regression models (Tibshirani, 1996). Other recent approaches include methods developed by Loh (2002) and Zeileis et al. (2008), which use significance tests to select splitting variables in subgroup identification. The activity region finder, proposed by Amaratunga et al. (2014), involves greedy tree search algorithms and the use of trinary splits (as opposed to conventional binary splits). A key feature of the approach is that it focuses on finding regions of high differential responses among groups of interest (Alvir et al., 2007, 2008). More recently, Foster et al. (2011) introduced a stepwise approach, called virtual twins, that involves predicting response probabilities for treatment and control in a counterfactual framework, which are then used as the outcome in a classification or regression tree.

Latent class analysis has also been applied to determine the presence and number of subgroups. In one study, for example, Clark et al. (2011) reported the results of a latent class analysis that showed the existence of subgroups of patients (classified as optimal responders, average responders, global responders, or nonresponders), characterized by distinct patterns of drug response using data from a large RCT.

In application, one should distinguish between confirmatory and exploratory subgroup analyses. While the latter is intended to generate hypotheses about the risks and benefits of a treatment option in a subset of the target population, the former is intended to draw definitive conclusions and requires prespecification of study objectives and analytical strategies.

The medical literature is now rich with examples of interest in the study of patients' demographic and clinical characteristics as potential predictive factors for drug effect. For example, Assmann et al. (2000) reported that 70% of the studies, which were published between July and September of 1997 in selected medical journals, included at least one subgroup analysis. In view of the growing interest, best practices have been proposed to ensure the reliability and integrity of such analyses (Wang et al., 2007).

7.4 PROs in Precision Medicine

A clinical outcome assessment (COA) directly or indirectly measures how patients feel or function and, hence, can be used to determine whether a treatment has demonstrated efficacy or effectiveness, as well as safety (Cappelleri and Spielberg, 2015; Walton et al., 2015). A COA measures a specific concept (i.e., what is being measured by an assessment, such as pain intensity) within a particular context of use. A COA is different from a biomarker assessment. Relative to COA, a biomarker assessment is one that is subject to less influence from either the patient or another rater on a patient's assessment.

There are four types of COAs: PROs (the patient as a rater), clinician-reported outcomes (a clinician as a rater), observer-reported outcomes (someone other than the patient or clinician, who may not have specialized professional health care training, as a rater), and performance outcome assessments (where the patient is instructed to perform a defined task in a specific objective way, such as distance walked in six minutes). In this section, the focus is on PROs, although what is stated about them may also be applicable to the other three COAs.

PRO instruments are designed to quantify concepts pertaining to how a patient feels or functions, and to help generate evidence of a treatment benefit or harm from the patient's perspective. Accordingly, they can provide complementary information to other clinical endpoints for use in personalized medicine. Indeed, accumulating data from studies suggest the existence of association between the genetic disposition of patients and their health status or health-related quality of life (QOL). For example, in one study, Rijsdijk et al. (2003) noted that the overall heritability of psychosocial distress ranged from 20% to 44%. Other studies (e.g., Romeis et al., 2000) have also reported evidence of genetic influences on PROs.

As discussed elsewhere (Sprangers et al., 2010), the study of the genetic disposition of PROs requires a conceptual model to establish the relationships among QOL domains (or subscales), biological mechanisms, and genetic variants. In addition to generic PRO measures, which can be administered irrespective of the illness or condition of the patient, and may be applicable to healthy people as well, disease-specific PRO measures may capture certain aspects of health status found in the generic PROs, depending on their purpose, but are usually specifically tailored to measure the special characteristics found in particular conditions or diseases. As such, relative to a generic measure, a disease-specific measure is expected to be more closely associated with a biomarker for a particular disease.

In recent years, the role of QOL in personalized medicine has continued to gain increasing attention, owing in part to the activities of organizations that aim to promote research on biological mechanisms, potential genes, and genetic variants involved in QoL (Sprangers et al., 2009). Advances in the area include summaries on the genetic background of common symptoms and overall well-being (Sprangers et al., 2009). Considerable progress has also been made in specific areas, such as oncology, to quantify the association between polymorphisms and PROs (Yang et al., 2009). In one instance, to assist with future efforts to adopt QOL monitoring while acknowledging the importance of biomarkers, a systematic review was undertaken to establish which domains of QOL are most affected by end-stage renal disease, and to measure the strength of evidence linking common biomarkers to QOL in end-stage renal disease (Spiegel et al., 2008).

It is now widely recognized that PROs should be an integral component of the assessment of the risk-benefits of alternative treatment options. Accordingly, a PRO protocol must be an integral part of the overall research plan. Particular attention should also be paid to the determination of which PRO concept is important to assess in a given disease area. When possible, generally accepted measures, rather than individually developed PROs for a specific therapy, should be used within a therapeutic area. Standardized measures not only facilitate interpretation of results, but also allow synthesizing data from different studies.

Nevertheless, despite the prominent place PROs are now given in medical research, their role in personalized medicine remains not yet fully explored. Clearly, improved understanding of the genetic basis of PROs could potentially facilitate the customization of treatment for an individual patient, as PROs are inherently intended to extract the overall clinical condition of a patient first-hand, without the known drawbacks of secondary reports by a health care provider (Alemayehu and Cappelleri, 2012). Although there is encouraging evidence relating to the impact of genetics on QOL and PROs (Leinonen et al., 2005; Raat et al., 2010), the evaluation of the precise proportion of variability in PROs that is explained by genetic factors still needs further elucidation, with more aggressive research than it has garnered currently.

In the following subsections, we provide a high-level summary of pertinent aspects of PRO data collection, analysis, and reporting, with an emphasis on the issues that are germane to their effective use in personalized medicine and other areas. Interestingly enough, PRO and genomic studies share many overlapping analytical and conceptual issues, including establishing the reliability and validity of instruments and in the handling of multiplicity and missing data. For a more in-depth discussion of these issues, the readers are referred to Alemayehu and Cappelleri (2012, 2014).

7.4.1 Instrument Development and Validation

For a PRO instrument to serve as a valuable tool in evidence-based medicine, it has first to undergo a rigorous evaluation to establish its validity and reliability using both conceptual and empirical approaches (Cappelleri et al., 2013). The conceptual assessment is geared toward ensuring that the reported outcome is linked to the specific symptoms or other aspects being measured. At the initial stage in the development of an instrument, data from focus groups and cognitive interviews with patients are customarily used to establish content validity of an instrument, which involves measuring what patients consider to be the important outcomes of the condition or disease (see, e.g., Chapter 2 in this volume).

Rigor in the development of the content of a PRO instrument is essential to ensure that the concept of interest is measured accurately, to capture issues of most relevance to the patient, and to subscribe to a language that allows patients to understand and respond without confusion. The presence of items within a questionnaire with little relevance to the patient population being investigated or, if the items are poorly written, will lead to measurement error and bias. If the questions within a measure are neither understood properly nor very relevant, then an ambiguous response is likely. Therefore, taking time to talk with patients about their symptoms or the impact of a disease or condition on their health status is very important before embarking on the generation of questions to measure the concept of interest.

Qualitative research with patients is essential for establishing content validity of a PRO measure. Steps to consider for conducting qualitative research to establish content validity are found elsewhere (Patrick et al., 2011a,b). Content validity is the basis for aiding in the interpretation of scores and for providing clarity for communication of findings.

In the next phase of instrument development or modification, data from carefully planned studies involving the target patient populations are essential to establish the measurement properties of the PRO measure—evidence for its reliability and validity of the measure—using standard psychometric techniques, including classical test theory and item response theory (see also Chapter 2 of this monograph). Validity assesses the extent to which an instrument measures what it is meant to measure, while reliability assesses how precise or stable the instrument measures what it measures.

Construct validity involves the degree to which the scores of a measurement instrument are consistent with hypotheses. Assessments of construct validity make use of correlations, changes over time, and differences between groups of patients.

Two approaches of (construct) validity assessment are exploratory factor analysis and confirmatory factor analysis. In exploratory factor analysis, there is initial uncertainty as to the number of factors being measured, as well as regarding which items are representing those factors. As such, the technique is suitable for generating hypotheses about the structure of distinct concepts and which items represent a particular concept. In contrast, confirmatory factor analysis is a hypothesis-confirming technique that relies on a researcher's hypothesis, and that requires prespecification of all aspects of the factor model. While exploratory factor analysis explores the patterns in the correlations of items, confirmatory factor analysis tests whether the correlations conform to an anticipated or expected scale structure given in a particular research hypothesis.

It is noted here that item response theory is another way to assess validity. As described elsewhere (Cappelleri et al., 2013), it is a statistical theory consisting of mathematical models expressing the probability of a particular response to a scale item as a function of the (latent or unobserved) attribute of the person and of certain parameters or characteristics of the item.

As part of the validation of an instrument, two standard types of reliability for PRO measures used are internal reliability for multi-item scales and repeatability reliability. Both concepts are mathematically related. Internal reliability, also known as internal consistency, is based on item-to-item correlations and the number of items in multi-item scales. Repeatability reliability, which is applicable to single-item scales as well as to multi-item scales, is based upon the analysis of variances between repeated measurements on the same set of subjects, where the measurements are repeated over time for patients in a stable condition under practically identical circumstances (test-retest reliability).

Under the assumption that a PRO measure has demonstrated validity and related, subsequent analytic issues ensue. They are briefly discussed in what follows.

7.4.2 Longitudinal Models

Most PRO data in clinical studies are longitudinal in nature and, in order to characterize changes in the response of interest over time, longitudinal models should be analyzed using techniques that take into account the correlation induced by repeated measurements on the same individual over time. Another common feature of longitudinal data is heterogeneous variability; that is, the variance of the response changes over the duration of the study. These two features of longitudinal data—correlated and heterogeneous responses over time—violate the fundamental assumptions of independence and homogeneity of variance that are the basis of many standard techniques (e.g., *t*-test, analysis of variance, multiple linear regression).

Statistical models for longitudinal data account for these two features. While the main scientific interest is typically focused on modeling the mean response, inferences about change in response over time (within group and between groups) can be sensitive to the chosen model for the covariance among the repeated measures. Failure to properly account for the covariance can result in hypothesis tests and confidence intervals that may be invalid and subsequently result in misleading inferences.

Responses on PRO measures in clinical studies can be based on a balanced design or an unbalanced design. A design is "balanced" when all individuals have the same number of repeated measurements that occur at a common set of occasions. Otherwise, a design is said to be "unbalanced." Modern methods for longitudinal data can accommodate unbalanced data, as well as balanced data, and also missing or incomplete data on the response outcome for a given individual (to be discussed a little later in this section).

Approaches to modeling continuous longitudinal data include the analysis of response profiles and linear mixed-effects models (random coefficient models). A repeated measures model with time taken as a categorical covariate, also known as an analysis of response profiles, is most appropriate when the data are balanced and the sample size is large relative to the number of measurement occasions. When this is not the case, a repeated measures model with time taken as a continuous covariate can be considered. This model results in a growth curve model with simple parameters (e.g., linear or quadratic) or semi-parametric (e.g., piecewise linear) curves.

Alternatively, linear mixed-effect models (with time taken a continuous covariate) can be implemented. They too provide flexibility for the analysis of longitudinal data by accommodating unbalanced data, and allow for the mixture of discrete and continuous covariates. Modeling of the covariance among repeated measures can be applied with a relatively small number of parameters. The distinctive feature of linear mixed-effects models is that the mean response is modeled with a combination of population characteristics that are assumed to be shared by all individuals (a set of fixed effects) and subject-specific effects that are unique to a particular individuals (a set of random effects).

Other methods are available for longitudinal studies in which the response is not continuous, for example, studies with repeated binary measurements. Some of these methods can be conceptualized as extensions of linear mixed-effect models, referred to as generalized linear mixed effects, which accommodate different types of responses. Other methods for noncontinuous (and continuous) outcomes, referred to as generalized estimating equations, can also be considered. For further details about methods for longitudinal data and examples of their application, several books are available (Fitzmaurice et al., 2011; Hedeker and Gibbons, 2006; Singer and Willett, 2003), including books specific to the longitudinal analysis of PROs (Cappelleri et al., 2013; Fairclough, 2010).

7.4.3 Item Response Theory (IRT)

IRT is a statistical theory consisting of nonlinear logistic models to express the probability of a particular response to a scale item as a function of the quantitative attribute of interest (called a latent or "unobservable" attribute, like depression) (Cappelleri et al., 2013; de Ayala, 2005; Hambleton et al., 1991; Hays et al., 2000). The mathematical description for the item response is called an item characteristic curve, which gives the probability of responding to particular category of an item for an individual with an estimated amount on the attribute. Each item typically has its own level of difficulty, where items that are more difficult are harder to endorse. For example, "running" would be more difficult (harder to endorse) than "walking" in measuring the concept of physical functioning; patients would be more likely to respond "strongly disagree" to "running" than to "walking" (if the range options were "strongly agree," "agree," "neither agree nor disagree," "agree," and "strongly disagree").

In addition, each item can have its own level of discrimination (items with more discrimination are more likely to distinguish among persons with varying levels on the attribute), depending on whether a one-parameter model (with an item difficulty parameter, along with a person attribute parameter) or a two-parameter model (with item difficulty and discrimination parameters, along with a person attribute parameter), is fit.

Applications and relevance of IRT for PROs has increased considerably over the last several years. For instance, IRT has been the cornerstone of the Patient-Reported Outcomes Measurement Information System (PROMIS), a large initiative of the National Institute of Health (NIH), which aims to revolutionize the way PROs are selected and employed in clinical research and practice evaluation. The broad objectives of the NIH PROMIS network are to develop and test a large bank of items measuring PROs; create a computerized adaptive testing system that allows for an efficient, psychometrically robust assessment of PROs in clinical research involving a wide range of chronic diseases; and create a publicly available system that can be added to and modified periodically, and which allows clinical researchers to access a common repository of items and computerized adaptive tests (Cella et al., 2007a,b).

Computer adaptive testing (CAT) is ideally suited to IRT in a personalized medicine setting. Traditional, fixed-length tests require administering some items that are too high for those with low levels of an attribute (e.g., asking patients with an extremely poor level of physical function whether they can run around the block) and other items that are too low for those with high levels of an attribute (e.g., asking patients with an extremely high level of physical function whether they can walk at a leisurely pace). These types of items yield little information to target a patient's location on an attribute of interest. In contrast, CAT makes it possible to estimate patient attribute levels based on a targeted subset of items in an item pool—a subset that is personalized or specific to a particular patient.

As an example, consider an item bank of highly discriminating items of varying difficulty levels that has been developed. A CAT algorithm that will ask enough items to sufficiently derive a response score can be applied using the following steps:

1. If there is no prior information, the first item administered is a randomly selected item of medium difficulty.

2. If the patient did not respond favorably to the first item (e.g., did not endorse a binary item, or responded on the low end of a scale for an ordinal item), the second item administered is an easier item located a prespecified step away from the first item. If the patient did respond favorably to the first item, the second item administered is a more difficult item located the specified step away.

3. The process continues and, after one item has a favorable response and another item a not-so-favorable response, maximum likelihood scoring is possible. The scoring can begin with the next item selected being the one that maximizes the likelihood function (e.g., an item with a 50% chance of being endorsed).

4. Finally, the process is terminated and the patient's true attributed level is estimated when the standard error on the attribute level falls below the acceptable value.

7.4.4 Missing Data with PROs

Missing data are a common problem in PRO analysis and require considerable attention to ensure unbiased reporting of results. For a given patient, PRO data may be missing for an entire visit, or for certain components of the instrument (item-level missing). Many instruments include well-documented procedures by their developers on how to handle missing items, and such recommendations by developers are typically the preferred way to address missing items when some items have a nonresponse (Cappelleri et al., 2013).

The simplest solution to dealing with missing items on a multi-item PRO measure is to treat the scale score as missing. This approach, of course, results in a loss of power and the threat of serious bias when the missingness is informative regarding the patient's true PRO status. Single imputation methods, with varying degrees of complexity, exist for missing item-level data (Fairclough, 2010). These methods make strong assumptions about the missing data mechanism, which may be unrealistic and may induce a systematic underestimate of the sampling variation in certain circumstances. The Food and Drug Administration (FDA, 2009) guidance states that "the [statistical analysis plan (SAP)] can specify the proportion of items that can be missing before a domain is treated as missing." This approach requires rules that prespecify the number of items that can be missing to still consider the domain as adequately measured.

One typical approach is to prorate a scale score if at least half of the items from a PRO scale have been answered (Fairclough, 2010; Fairclough and Cella, 1996). If a subject completes at least half of the items being used to compute a multiitem scale or domain of a PRO measure, this one-half rule procedure of imputation takes the mean of a patient's observed responses and substitutes it for the values of the missing items on the same scale. The ensuing summated score (the simple sum of the observed and imputed values) is equivalent to taking the mean of the observed responses and multiplying by the total number of items.

A strength of the one-half rule method is that it imputes a person-specific estimate for each missing item, based on the average score across completed items on the same scale from the same person. Thus, unlike some other single imputation methods, the imputation of scores does not depend on data collected from other subjects, other trials, or even other measures from the same subject. The one-half rule method is well-suited for multiitem scales in which there is no clear ordering or hierarchy of item difficulty on the questions. Research has suggested that the approach, though simple, compares favorably (in terms of being most unbiased and precise) against regression-based and other single-imputation methods for item nonresponse (Fairclough and Cella, 1996).

Handling missing data is a complex matter, and there is no universally accepted approach to mitigate the impact on subsequent results. As a first line of defense, every attempt should be made to minimize the occurrence of missing data during the design and conduct phases of the study. For example, use of new technologies such as ePROs may help maximize PRO data quality and collection.

A more complex situation than item-level missing on a domain or questionnaire is data missing from all items on a domain or questionnaire. Appropriate measures in the design and conduct stages of clinical trials can help to limit missing data (Little et al., 2012a,b). At the design stage, the following eight ideas have been suggested: (1) target a population that is not adequately served by treatment, (2) include a run-in period, (3) allow for a flexible treatment regimen, (4) consider an add-on design, (5) shorten the follow-up period, (6) allow the use of rescue medications, (7) consider a randomized withdrawal design (for long-term efficacy), and (8) avoid outcome measures that are likely to lead to substantive missing data.

At the conduct stage, the following eight ideas have been suggested: (1) select investigators with a good record for collecting complete data, (2) set acceptable target rates for missing rates and monitor the progress made, (3) provide monetary and nonmonetary incentives to investigators and participants, (4) limit the burden and inconvenience of data collection, (5) provide continued access to effective treatments after the trial (before treatment approval), (6) train investigators and study staff on ways to keep participants in the study until the end, (7) collect information from participants regarding likelihood that they will drop out, and (8) keep contact information for participants up-to-date. The relevance of the suggestions at the design and conduct stages varies greatly according to setting, and they may have limitations or drawbacks that need to be considered.

At the analysis stage, alternative approaches may be used depending on the missingness mechanism. When data are missing at random (MAR), multiple imputation (Carpenter and Kenward, 2013) and the mixed effects model with repeated measures (Fitzmaurice et al., 2011) are routinely used. In certain situations, the missingness mechanism may be obvious by inspecting the study design or other aspects of the study conduct. However, in general, it may not be possible to ascertain whether the assumption of MAR can be justified. As a best practice, it is therefore advisable to perform sensitivity analysis to ensure the robustness of the findings under alternate scenarios (Little and Rubin, 2002; National Research Council, 2010). Pattern mixture models are sometimes used in sensitivity analysis, with missing values imputed under a plausible scenario in which the missing data are considered missing not at random (MNAR) (Little, 1993). These techniques are now readily implemented in the major statistical software packages, such as SAS (SAS Institute, Cary, NC) with its MI procedure, by using the MNAR statement. An alternative approach, which is relatively less dependent on assumptions, involves searching for a tipping point that reverses the study conclusion (O'Kelly and Ratitch, 2014).

7.4.5 Interpretation

Often, the findings from a PRO analysis may not be readily interpretable. Accordingly, two approaches—an anchor-based approach and a distribution-based approach—have been used to aid in the understanding of the relevance of PRO scores (Cappelleri and Bushmakin, 2014; Marquis et al., 2004; McLeod et al., 2011). Anchor-based approaches are the preferred way to enhance the clinical interpretation to the targeted PRO measure. An anchor-based approach links the targeted concept of the PRO measure to the meaningful concept (or criterion) emanating from the anchor, such as patient assessment on the severity of the condition.

The approach maps the PRO under consideration with an anchor measure or indicator that is interpretable itself or lends itself to interpretation. The anchor should meet at least two criteria: (1) be appreciably correlated with the targeted PRO (say, a correlation between 0.4 and 0.7); and (2) be easy to interpret and certainly more easy to interpret than the PRO itself. The anchor may or may not be another PRO measure, although it is preferred that it is such when the target measure is a PRO. Considerations for anchor-based methods include the nature of the relationship (e.g., linear) between the anchor and the targeted PRO, the type of anchor, and the study population of interest. Examples of anchor-based methods include percentages based on thresholds, criterion-group interpretation, content-based interpretation, and regression-based methods for determining clinically important differences (Cappelleri and Bushmakin, 2014; Cappelleri et al., 2013).

Distribution-based approaches are also used to evaluate the magnitude of a treatment effect, both at the individual and group levels (Alemayehu and Cappelleri, 2012). Distribution-based methods can offer valuable insights about the magnitude of an effect in terms of a signal-to-noise ratio. These methods also allow for a standardization of different scales with different ranges and ways of scoring. On the other hand, a limitation of distribution-based methods should be noted: although their interpretation can be considered meaningful, they do not provide information about clinical meaningfulness. Examples of distributed-based approaches for a group of patients include effect size, probability of relative benefit, responder analysis, and cumulative distribution functions (Cappelleri et al., 2013; Cappelleri and Bushmakin, 2014).

Group-level change can be statistically significant but trivial in magnitude if the group sample size is large enough. With individual-level change, the size of differences required to be statistically significant will not be trivial. For change in individual patients, several metrics are available to assess the statistically significant change in PRO measures (Hays et al., 2005).

7.5 Regulatory, Access, and Other Considerations in Precision Medicine

7.5.1 HTA

In many countries, HTA is the primary tool for making decisions regarding coverage decisions for drugs, medical devices, and health care procedures. In the context of personalized medicine, the process tends to be arduous, lacking best practices for value assessment. First, an adequate assessment of the clinical benefits is limited because of the small samples studied in trials, which were not typically designed to address issues of precision medicine. In addition, there is often insufficient economic evidence due to the paucity of data on clinical utility, particularly prior to adoption. In many cases, comparators are often imprecisely defined. As a consequence, one may not be able to estimate incremental cost-effectiveness ratios with adequate precision.

In view of the difficulties that are pervasive in personalized medicine, it has been proposed that new evaluation methods for HTA be developed (Meckley and Neumann, 2010). One suggested approach is to use the framework already available for orphan drugs and other treatments intended for rare diseases. Effective operationalization of such an approach would, nonetheless, presuppose close collaboration among concerned stakeholders, including pharmaceutical companies, regulators, and HTA agencies.

Despite the many challenges, there are a few examples in which drug developers have been able to successfully garner payer acceptance (Faulkner et al., 2012). In most cases, a major factor in ensuring optimal personalized medicine market access has been the availability of companion diagnostic or biomarker tests with adequate evidence to address HTA agency and payer needs (Spinner et al., 2013). A case in point is trastuzumab therapy, which illustrates a situation where the driving factor for success was the availability of the companion diagnostics at the time of launch (Ferrusi et al., 2009). There are also situations where reimbursement decisions were delayed until a companion diagnostic was available, as was the case with gefitinib (PBAC, 2009). However, acceptance is not always guaranteed, especially when the evidence supporting the cost-benefit is not of high quality (Paci and Ibarreta, 2009). In other cases in which the diagnostic tool and the medicines are not concurrently developed, it may not be feasible to conclusively establish the relationships among the diagnostic, treatment, and health outcomes, resulting in lack of acceptance by payers (Centers for Medicare & Medicaid Services, 2010; The International Warfarin Pharmacogenetics Consortium, 2009).

7.5.2 Regulatory Issues

As in rare diseases, precision medicine development typically involves important operational and ethical issues, including dealing with the small size of the study populations, and conducting trials in vulnerable groups. Accordingly, the existing paradigm for the licensing of pharmaceutical products, which is characterized by different phases of development and several well-designed RCTs, needs to be revamped to be in harmony with the constraints imposed by precision medicine reality.

There are steps that may be taken to mitigate some of the problems with precision medicine development. For example, the use of novel approaches, including adaptive study designs, seamless Phase 2/3 programs, and Bayesian analytical techniques could substantially accelerate the development timelines (see, e.g., Schmidli et al., 2006). Further, when trials based on a biomarker design are not feasible (e.g., when a biomarker has a low prevalence), other novel strategies, such as patient-centric development, could be explored. In this regard, molecularly informed trials appear to be promising. In oncology drug development, for example, basket studies can be used to study the effects of a single drug on a particular mutation that is associated with different cancer types (Redig and Jänne, 2015). Additionally, when interest lies in studying the effects of different drugs on different mutations in a single type of cancer, an umbrella trial design may be employed (see, e.g., Kim et al., 2011). There is also a growing interest in n-of-1 studies, which focus on exceptional responders (Schork, 2015).

While there are currently no specific guidelines for the approval of precision medicines, other than relying on existing pathways intended to accelerate approvals for certain classes of drugs, recent measures, such as the adaptive licensing proposal in the European Union (EU) and the progressive licensing framework in Canada, appear to be steps in the right direction. Further, with regards to prognostic tests, the regulatory guidelines in both the US and the EU appear to require greater harmonization. For example, although the FDA requires the establishment of both the analytic validity and clinical validity of commercial diagnostic tests, there is no formal requirement for certain other genomic tests. On the other hand, in the EU, commercial companion diagnostics, treated as *in vitro* diagnostics, are only subjected to self-certification by the manufacturer. The recent activities by the FDA (2013), including the report on how the agency is changing its regulatory infrastructure to facilitate personalized medicine, may help address some of the issues (Evans et al., 2015).

7.5.3 Ethical Issues

The effective implementation of personalized medicine in clinical practice requires the institution of measures to safeguard the privacy of individuals and their ability to access health care resources. Improper use and dissemination of genomic data could adversely compromise the privacy of patients, with far-reaching consequences, including genetic stigmatization, economic deprivation, and denial of insurance coverage. While there are ongoing initiatives in the legislative and ethical arenas to ensure the protection of privacy and confidentiality, much work is needed to warrant the proper use of precision medicine to advance public health (Lee and Mudaliar, 2009).

7.5.4 Clinical Practice

A study was conducted to elicit the preferences of physicians regarding applying personalized medicine in their clinical practice as these strategies become available (Najafzadeh et al., 2012). Najafzadeh et al. (2012) reported a best-worse scaling choice experiment, conducted to estimate the relative importance of attributes that influence physicians' decisions for using personalized medicine.

Six attributes emerged: type of genetic tests, training for genetic testing, clinical guidelines, professional fee, privacy protection laws, and cost of genetic tests. "Type of genetic tests" had the largest importance, suggesting that the physician's decision was highly influenced by the availability of genetic tests for patients' predisposition to disease or drug response. "Training" and "guidelines" were attributes with the next highest importance. Latent class analysis identified two classes of physicians. Relative to class 2, class 1 had a larger weight for a "type of genetic test" and smaller weights for "professional fee" and "cost of tests." These results can be used to design the policies for supporting physicians and facilitating the use of personalized medicine in the future.

7.6 Concluding Remarks

The concept of identifying and targeting patients that respond to a given treatment appears to be promising to advance health care by improving overall effectiveness while reducing costs. However, this paradigm would be more economically viable if the cost of establishing the screening process to personalize treatment does not outweigh the accrued savings. For example, the development of reliable and valid biomarkers is an essential element in optimizing the economic value of personalized medicines. While there is hope that the increased use of next-generation sequencing may be more cost-effective than current biomarker tests, their implementation in clinical practice is dependent on the availability of data on the clinical utility of such tests.

With the growing interest in personalized medicine, there are compelling reasons to incorporate PROs as an integral part of the research endeavor in personalized medicine (Alemayehu and Cappelleri, 2012). Specifically, insights into the genetics of PROs will ultimately allow for the early identification of patients susceptible to PRO deficits, as well as the targeting of care in advance. Therefore, by unraveling the genetic understings of PROs (e.g., what specific single-nucleotide polymorphisms, on which specific genes, are associated with pain), researchers will have a greater understanding of diagnosis and treatment management for an individual patient, an understanding that has the potential to lead to improved survival, PRO assessments, and health service delivery. However, to ensure that PROs play an effective complementary role to traditional clinical endpoints in personalized medicine, it is essential to understand the issues that are inherent in PRO data, and to put in place processes to guide researchers and other stakeholders.

We highlighted the need for a conceptual frame to incorporate PRO data in personalized medicine and reviewed methodological and analytical approaches that are relevant for the analysis and interpretation of PROs. This provides challenges and opportunities from the standpoint of application, as well as for methodological research. Recent developments in the specialized areas of PROs and personalized medicine provide fertile opportunities to bring the two areas even closer, and to advance the way treatment is attuned and delivered to address patient care and needs.

Personalized medicine and PROs have attracted considerable attention from regulatory agencies. For example, the US FDA provides a roadmap for the inclusion of PROs in a label claim (FDA, 2009). Similar efforts are also under way to establish the regulatory science for evaluating the strategies and outcomes for personalized medicine (FDA, 2013). Nonetheless, the current paradigm of drug development and approval is not particularly suited to fully realize the promise of precision medicine. As the technology for gene sequencing advances and our understanding of the interplay among genes, and that between genes and the environment, grows, it is imperative to bring to bear the concerted efforts of all stakeholders, including academia, drug developers, HTA agencies, and regulatory bodies, to positively transform the health care delivery ecosystem.

Acknowledgments

Special thanks go to David Gruben for helpful comments. Some material relating to PROs was taken from Cappelleri et al. (2013).

References

Alemayehu, D. and J.C. Cappelleri. 2012. Conceptual and analytical considerations toward the use of patient-reported outcomes in personalized medicine. *Am Health Drug Benefits* 5:310–317.

Alemayehu, D. and J.C. Cappelleri. 2014. Patient-reported outcomes in personalized medicine. In *Clinical and Statistical Considerations in Personalized Medicine*, ed. C. Carini, S. M. Menon, and M. Chang, pp. 297–312. Boca Raton, FL: Chapman and Hall/CRC.

Alvir, J., Cabrera, J., Caridi, F. et al. 2007. Mining clinical trial data. In *Knowledge Discovery and Data Mining: Challenges and Realities*, ed. X. Zhu and I. Davidson, pp. 31–42. Hershey, PA: Information Science Reference (IGI Global).

Alvir, J., Cabrera, J, Caridi, F. et al. 2008. A robust recursive partitioning algorithm for mining multiple populations. In *Frontiers of Applied and Computational Mathematics: Dedicated to Daljit Singh Ahluwalia on his 75th Birthday*, ed. D. Blackmore, A. Bose, and P. Petropoulos, pp. 75–82. Hackensack, NJ: World Scientific.

Amaratunga, D., Cabrera, J. and Y.S. Lee. 2008. Enriched random forests. *Bioinformatics* 24:2010–2014.

Amaratunga, D., Cabrera, J. and Z. Shkedy. 2014. *Exploration and Analysis of DNA Microarray and Other High-Dimensional Data*, 2nd edition. Hoboken, NJ: John Wiley & Sons.

Assmann, S.F., Pocock, S.J., Enos, L.E. et al. 2000. Subgroup analysis and other (mis) uses of baseline data in clinical trials. *Lancet* 355:1064–1069.

Benjamini, Y. and Y. Hochberg. 1995. Controlling the false discovery rate: A practical and powerful approach to multiple testing. *J Roy Stat Soc B* 57:289–300.

Breiman, L. 2001. Random forests. *Mach Learn* 45:5–32.

Cappelleri, J.C. and A.G. Bushmakin. 2014. Interpretation of patient-reported outcomes. *Stat Methods Med Res* 23:460–483.

Cappelleri, J.C. and S.P. Spielberg. 2015. Advances in clinical outcome assessments. *Ther Innov Regul Sci* 49:780–782.

Cappelleri, J.C., Zou, K.H., Bushmakin, A.G. et al. 2013. *Patient-Reported Outcomes: Measurement, Implementation and Interpretation*. Boca Raton, FL: Chapman & Hall/CRC Press.

Carpenter, J. R. and M.G. Kenward. 2013. *Multiple Imputation and Its Application*. New York, NY: John Wiley & Sons.

Cella, D., Gershon, R.C., Laiet, J.S. et al. 2007a. The future of outcomes measurement: Item banking, tailored short-forms, and computerized adaptive assessment. *Qual Life Res* 16(Suppl. 1):133–141.

Cella, D., Yount, S., Rothrock, N. et al. 2007b. The patient-reported outcomes measurement information system (PROMIS): Progress of an NIH Roadmap cooperative group during its first two years. *Med Care* 45(5 Suppl. 1):S3–S11.

Centers for Medicare & Medicaid Services. 2010. *National Coverage Determination (NCD) for pharmacogenomic testing for warfarin response (90.1)*. http://www.cms .gov/medicare-coverage-database/details/ncd-details.aspx?NCDId=333&ncd ver=1&bc=BAAAgAAAAAAA& (accessed March 20, 2016).

Chau, C.H., Rixe, O., McLeod, H. et al. 2008. Validation of analytic methods for biomarkers used in drug development. *Clin Cancer Res* 14:5967–5976.

Clark, S.L., Adkins, D.E. and E.J.C.G. van den Oord. 2011. Analysis of efficacy and side effects in CATIE demonstrates drug response subgroups and potential for personalized medicine. *Schizophr Res* 132:114–120.

Couch, R.D. and B.T. Mott. 2012. Personalized medicine: Changing the paradigm of drug development. *Mol Biol Methods and Protocols*, 823:367–378.

de Ayala, R.J. 2005. *The Theory and Practice of Item Response Theory*. New York, NY: Guilford Press.

DeRubeis, R.J., Cohen, Z.D., Forand, N.R. et al. 2014. The personalized advantage index: Translating research on prediction into individualized treatment recommendations. *PLOS ONE* 9(1):Article ID e83875. http://journals.plos.org/ plosone/article?id=10.1371/journal.pone.0083875 (accessed May 24, 2017).

Deverka, P.A., Vernon, J. and H.L. McLeod. 2010. Economic opportunities and challenges for pharmacogenomics. *Annu Rev Pharmacol Toxicol* 50:423–437.

Drummond, M.F., Sculpher, M.J., Torrance, G.W. et al. 2005. *Methods for the Economic Evaluation of Health Care Programmes*, 3rd edition. New York, NY: Oxford University Press.

Evans, B.J., Burke, W. and G. Jarvik. 2015. The FDA and genomic tests—Getting the regulation right. *N Eng J Med* 372:2258–2264.

Fairclough, D.L. 2010. *Design and Analysis of Quality of Life Studies in Clinical Trials*, 2nd edition. Boca Raton, FL: Chapman & Hall/CRC Press.

Fairclough, D.L. and D.F. Cella. 1996. Functional Assessment of Cancer Therapy (FACT-G): Non-response to individual questions. *Qual Life Res* 5:321–329.

Faulkner, E., Annemans, L., Garrison, L. et al. 2012. Challenges in the development and reimbursement of personalized medicine—Payer and manufacturer perspectives and implications for health economics and outcomes research: A report of the ISPOR personalized medicine special interest group. *Value Health* 15:1162–1171.

Ferrusi, I.L., Marshall, D.A., Kulin, N.A. et al. 2009. Looking back at 10 years of trastuzumab therapy: What is the role of HER2 testing? A systematic review of health economic analyses. *Per Med* 6:193–215.

Fitzmaurice, G.M., Laird, N.M. and J.H. Ware. 2011. *Applied Longitudinal Analysis*, 2nd edition. Hoboken, NJ: John Wiley & Sons.

Food and Drug Administration (FDA). 2009. Guidance for industry on patient-reported outcome measures: Use in medical product development to support labeling claims. *Fed Regist* 74(235):65132–65133.

Food and Drug Administration (FDA). 2013. *Paving the way for personalized medicine:FDA's role in a New Era of medical product development*. http://www.fda .gov/downloads/ScienceResearch/SpecialTopics/PersonalizedMedicine/ UCM372421.pdf (accessed on April 22, 2016).

Foster, J.C., Taylor, J.M.C. and S.J. Ruberg. 2011. Subgroup identification from randomized clinical trial data. *Stat Med* 30:2867–2880.

Foster, M.W., Mulvihill, J.J. and R.R. Sharp. 2009. Evaluating the utility of personal genomic information. *Genet Med* 11:570–574.

Fragoulakis, V., Mitropoulou, C., Williams, M.S. et al. 2015. *Economic Evaluation in Genomic Medicine*. Boston, MA: Academic Press.

Gosho, M., Nagashima, K. and Y. Sato. 2012. Study designs and statistical analyses for biomarker research. *Sensors* 12:8966–8986.

Hambleton, R.K., Swaminathan, H. and H.J. Rogers. 1991. *Fundamentals of Item Response Theory*. Newbury Park, CA: Sage Publications.

Hays, R.D., Brodsky, M., Johnston, M.F. et al. 2005. Evaluating the statistical significance of health-related quality-of-life change in individual patients. *Eval Health Prof* 28:160–171.

Hays, R.D., Morales, L.S. and S.P. Reise. 2000. Item response theory and health outcomes measurement in the 21st century. *Med Care* 38:II-28–II-42.

Hedeker, D. and R.D. Gibbons. 2006. *Longitudinal Data Analysis*. Hoboken, NJ: John Wiley & Sons.

Holtzman, N.A. and M.S. Watson, eds. 1997. Promoting safe and effective genetic testing in the United States. *Final report of the task force on genetic testing*. http:// www.genome.gov/10001733 (accessed July 6, 2016).

The International Warfarin Pharmacogenetics Consortium. 2009. Estimation of the warfarin dose with clinical and pharmacogenetic data. *N Eng J Med* 360:753–764.

Kim, E.S., Herbst, R.S., Wistuba, I.I. et al. 2011. The BATTLE trial: Personalizing therapy for lung cancer. *Cancer Discov* 1:44–53.

Kostis, W., Cabrera, J., Hooper, W. et al. 2013. Relationships between selected gene polymorphisms and blood pressure sensitivity to weight loss in elderly persons with hypertension. *Hypertension* 61:857–863.

Lee, S.S. and A. Mudaliar. 2009. Racing forward: The genomics and personalized medicine act. *Science* 323:342.

Leinonen, R., Kaprio, J., Jylhä, M. et al. 2005. Genetic influences underlying self-rated health in older female twins. *J Am Geriatr Soc* 53:1002–1007.

Little, R.J.A. 1993. Pattern-mixture models for multivariate incomplete data. *J Am Stat Assoc* 88:125–134.

Little, R.J.A and D.B. Rubin. 2002. *Statistical Analysis with Missing Data*, 2nd edition. Hoboken, NJ: John Wiley and Sons.

Little, R.J., Cohen, M.L., Dickersin, K. et al. 2012b. The design and conduct of clinical trials to limit missing data. *Stat Med* 31:3433–3443.

Little, R.J., D'Agostino, R., Cohen, M.L. et al. 2012a. The prevention and treatment of missing data in clinical trials. *N Eng J Med* 367:1355–1360.

Loh, W.Y. 2002. Regression trees with unbiased variable selection and interaction detection. *Stat Sin* 12:361–386.

Lu, W., Zhang, H.H. and D. Zeng. 2013. Variable selection for optimal treatment decision. *Stat Methods Med Res* 22:493–504.

Ma, J., Hobbs, B.P. and F.C. Stingo. 2015. Statistical methods for establishing personalized treatment rules in oncology. *Biomed Res Int* 2015:670691. http://www.hindawi.com/journals/bmri/2015/670691/ (accessed July 6, 2016).

Marquis, P., Chassany, O. and L. Abetz. 2004. A comprehensive strategy for the interpretation of quality-of-life data based on existing methods. *Value Health* 7:93–104.

Matsui, S. 2013. Genomic biomarkers for personalized medicine: Development and validation in clinical studies. *Comput Math Methods Med* 2013:865980. http://www.ncbi.nlm.nih.gov/pmc/articles/PMC3652056/ (accessed July 6, 2016).

McLeod, L.D., Coon, C.D., Martin, S.A. et al. 2011. Interpreting patient-reported outcome results: US FDA guidance and emerging methods. *Expert Rev Pharmacoecon Outcomes Res* 11:163–169.

Meckley, L.M. and P.J. Neumann. 2010. Personalized medicine: Factor influencing reimbursement. *Health Policy* 94:91–100.

Molinaro, A.M., Simon, R. and R.M. Pfeiffer. 2005. Prediction error estimation: A comparison of resampling methods. *Bioinformatics* 21:3301–3307.

Najafzadeh, M., Lynd, L.D., Davis, J.C. et al. 2012. Barriers to integrating personalized medicine into clinical practice: A best-worst scaling choice experiment, *Genet Med* 14:520–526.

National Research Council: Committee on a Framework for Developing a New Taxonomy of Disease. 2011. *Toward Precision Medicine: Building a Knowledge Network for Biomedical Research and a New Taxonomy of Disease*. Washington, DC: The National Academies Press.

National Research Council: Panel on Handling Missing Data in Clinical Trials. 2010. *The Prevention and Treatment of Missing Data in Clinical Trials*. Washington, DC: The National Academies Press.

O'Brien, B.J. and A.H. Briggs. 2002. Analysis of uncertainty in health care cost-effectiveness studies: An introduction to statistical issues and methods. *Stat Methods Med Res* 11:455–468.

O'Donnell, J.C. 2013. Personalized medicine and the role of health economics and outcomes research: Issues, applications, emerging trends, and future research. *Value Health* 16:S1–S3.

O'Kelly, M. and B. Ratitch. 2014. *Clinical Trials with Missing Data: A Guide for Practitioners*. Hoboken, NJ: John Wiley and Sons.

Paci, D. and D. Ibarreta. 2009. Economic and cost-effectiveness considerations for pharmacogenetics tests: An integral part of translational research and innovation uptake in personalized medicine. *Current Pharmacogenomics Person Med* 7:284–296.

Patrick, D.L., Burke, L.B., Gwaltney, C.H. et al. 2011a. Content validity—Establishing and reporting the evidence in newly developed patient reported outcomes (PRO) instruments for medical product evaluation: ISPOR PRO good research practices task force report: Part 1—Eliciting concepts for a new PRO instrument. *Value Health* 14:967–977.

Patrick, D.L., Burke, L.B., Gwaltney, C.H. et al. 2011b. Content validity—Establishing and reporting the evidence in newly developed patient reported outcomes (PRO) instruments for medical product evaluation: ISPOR PRO good research practices task force report: Part 2—Assessing respondent understanding. *Value Health* 14:978–988.

Pepe, M.S., Longton, G., Anderson, G.L. et al. 2003. Selecting differentially expressed genes from microarray experiments. *Biometrics* 59:133–142.

Pharmacy Benefits Advisory Committee (PBAC). November 2009. *PBAC outcomes—Positive recommendations*. http://www.pbs.gov.au/info/industry/listing/elements/pbac-meetings/pbac-outcomes/2009-11/a-positive-recommend (accessed May 24, 2017).

Raat, H., van Rossem, L., Jaddoe, V.W. et al. 2010. The Generation R study: A candidate gene study and genome-wide association study (GWAS) on health-related quality of life (HRQOL) of mothers and young children. *Qual Life Res* 19:1439–1446.

Redig, A.J. and P.A. Jänne. 2015. Basket trials and the evolution of clinical trial design in an era of genomic medicine. *J Clin Oncol* 33:975–977.

Rijsdijk, F.V., Snieder, H., Ormel, J. et al. 2003. Genetic and environmental influences on psychological distress in the population: General health questionnaire analysis in UK twins. *Psychol Med* 33:793–801.

Romeis, J.C., Scherrer, J.F., Xian, H. et al. 2000. Heritability of self-reported health. *Health Serv Res* 35:995–1010.

Schmidli, H., Bretz, F., Racine, A. et al. 2006. Confirmatory seamless phase II/III clinical trials with hypothesis selection at interim: Applications and practical considerations. *Biometrical J* 48:635–643.

Schork, N. J. 2015. Time for one-person trials. *Nature* 520:609–611.

Simon, R. 2010. Clinical trial designs for evaluating the medical utility of prognostic and predictive biomarkers in oncology. *Per Med* 7:33–47.

Singer, J.D. and J.B. Willett. 2003. *Applied Longitudinal Data Analysis: Modeling Change and Event Occurrence*. New York, NY: Oxford University Press.

Soria, J.C., Blay, J.Y., Spano, J.P. et al. 2011. Added value of molecular targeted agents in oncology. *Ann Oncol* 22:1703–1716.

Speed, T. 2003. *Statistical Analysis of Gene Expression Microarray Data*. Boca Raton, FL: Chapman and Hall/CRC Press.

Spiegel, B.M.R., Melmed, G., Robbins, S. et al. 2008. Biomarkers and health-related quality of life in end-stage renal disease: A systematic review. *Clin J Am Soc Nephrol* 3:1759–1768.

Spinner, D.S., Ransom, J.F., Culp, J.L. et al. 2013. Health technology assessment of companion diagnostic biomarkers as gatekeepers for personalized medicine market access. *Value Health* 16:A376.

Sprangers, M.A., Sloan, J.A., Barsevick, A. et al. 2010. The GENEQOL Consortium. Scientific imperatives, clinical implications, and theoretical underpinnings for the investigation of the relationship between genetic variables and patient reported quality-of-life outcomes. *Qual Life Res* 19:1395–1403.

Sprangers, M.A.G., Sloan J.A., Veenhoven, R. et al. 2009. The establishment of the GENEQOL consortium to investigate the genetic disposition of patient-reported quality-of-life outcomes. *Twin Res Hum Genet* 12:301–311.

Storey, J. D. and R. Tibshirani. 2001. Estimating false discovery rates under dependence, with applications to DNA microarrays. *Technical report of the Stanford University Department of Statistics*. http://www.genomine.org/papers/dep.pdf (accessed May 24, 2017).

Storey, J.D. and R. Tibshirani. 2003. Statistical significance for genome-wide studies. *P Natl Acad Scie USA* 100:9440–9445.

Tibshirani, R. 1996. Regression shrinkage and selection via the lasso. *J Roy Stat Soc B* 58:267–288.

Trevor, H., Tibshirani, R. and J. Friedman. 2009. *The Elements of Statistical Learning: Data Mining, Inference and Prediction*, 2nd edition. New York, NY: Springer.

Walton, M.K., Powers, J.O.H., Hobart, J. et al. 2015. Clinical outcome assessments: Conceptual foundation – Report of the ISPOR clinical outcomes assessment—Emerging good practices for outcomes research task force. *Value Health* 18:741–752.

Wang, R., Lagakos, S.W., Ware, J.H. et al. 2007. Statistics in medicine—Reporting of subgroup analyses in clinical trials. *N Engl J Med.* 357:2189–2194.

Yang, P., Mandrekar, S.J., Hillman, S.H. et al. 2009. Evaluation of glutathione metabolic genes on outcomes in advanced non-small cell lung cancer patients after initial treatment with platinum-based chemotherapy: An NCCTG-97–24–51 based study. *J Thorac Oncol* 12:479–485.

Zeileis, A., Hothorn, T. and K. Hornik. 2008. Model-based recursive partitioning. *J Comput Graph Stat* 17:492–514.

Zou, K.H., Liu, A., Bandos, A.I. et al. 2011. *Statistical Evaluation of Diagnostic Performance: Topics in ROC Analysis*. Boca Raton, FL: Chapman and Hall/CRC Press.

8

Best Practices for Conducting and Reporting Health Economics and Outcomes Research

Kelly H. Zou, Joseph C. Cappelleri, Christine L. Baker, and Eric C. Yan

CONTENTS

8.1 Introduction 177
8.2 Guidelines on Patient-Reported Outcomes and Other Clinical
    Outcomes Assessments 178
8.3 Other Best Practices for HEOR 179
8.4 Concluding Remarks 180
References 180

8.1 Introduction

As presented in earlier chapters of this monograph, data from health economics and outcomes research (HEOR) is increasingly used to assess and enhance the effectiveness and efficiency of health care systems. In addition to randomized controlled trials (RCTs), HEOR utilizes data from other sources, including pragmatic trials and observational studies (Ford and Norrie, 2016). In recent years, observational data have especially gained wider acceptance for informing policies to improve patient outcomes and advise health technology assessment (AHRQ, 2013; Alemayehu and Berger, 2016; Berger and Doban, 2014; Cohen et al., 2015; Holtorf et al., 2012; Vandenbroucke et al., 2007; Zikopoulos et al., 2012).

In view of the complexity and heterogeneity of the data sources used in HEOR, it is important that research findings based on these outcomes be interpreted with caution, especially when considered possibly useful to support reimbursement decisions and also labeling claims, which must be consistent with medical product labels. In this chapter, we highlight a few guidance resources that are formulated as best practices in the conduct and reporting of HEOR studies.

8.2 Guidelines on Patient-Reported Outcomes and Other Clinical Outcomes Assessments

As indicated in earlier chapters (especially in Chapter 2), a patient-reported outcome (PRO) is any report on the status of a patient's health condition that comes directly from the patient, without interpretation of the patient's response by a clinician or anyone else (FDA, 2009). Generally, findings measured by PROs may be used to support claims in approved medical product labeling, if the claims on PROs are derived from adequate and well-controlled investigations. Furthermore, the PROs can measure specific concepts accurately and as intended, given the context of use.

Several regulatory guidelines are available relating to the development and assessment of PROs (EMA, 2005, 2014; FDA, 2009, 2014). The International Society for Pharmacoeconomics and Outcomes Research (ISPOR) PRO Good Research Practices Task Force (Patrick et al., 2011a,b) guideline addresses the conduct of qualitative research with patients that is essential for establishing content validity of a measure, especially for a label claim. This is a critical step because it is important to develop the concept of interest and its related question before embarking on full psychometric validation. By doing so, content validity will lay the framework to subsequently invite psychometric validation, which in turn aids in the interpretation of scores and in clarity for the communication of findings.

Calvert et al. (2013) gave additional guidelines for the reporting of PROs in RCTs. Five checklist items are recommended for RCTs in which PROs are primary or important secondary endpoints: (1) PROs should be identified as a primary or secondary outcome in the abstract, (2) a description of the hypothesis of the PROs and relevant domains should be provided, (3) evidence of the PRO instrument's validity and reliability should be provided or cited, (4) statistical approaches for dealing with missing data should be explicitly stated, and (5) PRO-specific limitations of study findings and the generalizability of the results to other populations and to clinical practice should be discussed.

In addition to PROs, clinical outcome assessments also include clinician-reported outcomes, observer-reported outcomes, and performance-based outcomes measures (Cappelleri and Spielberg, 2015). ISPOR has produced emerging good practices on clinician-reported outcome assessments (Cappelleri et al., 2017; Powers et al., 2017). More generally, the guidelines issued by the United States (US) Food and Drug Association (FDA) (2016) provide a roadmap for the four types of clinical outcome assessments.

8.3 Other Best Practices for HEOR

According to the Agency for Healthcare Research and Quality (AHRQ, 2017), comparative effectiveness research (CER) is designed "to inform health care decisions by providing evidence on the effectiveness, benefits, and harms of different treatment options. The evidence is generated from research studies that compare drugs, medical devices, tests, surgeries, or ways to deliver health care." Willke and Mullins (2011) have recommended "ten command-ments" be considered when conducting CER. These 10 important areas address study design, research questions, data sources, statistical analysis plan, and the interpretation of results.

Furthermore, ISPOR has sponsored an impressive list of best practices (Berger et al., 2009, 2012, 2014; Caro et al., 2014; Cox et al., 2009; Johnson et al., 2009). The ISPOR Good Practices for Outcomes Research Reports (ISPOR, 2016) provides expert consensus guidance recommendations on a wide range of topics for clinical and economic purposes, including PROs, for use in health care decisions. For example, guidelines for indirect comparison recognize that the best evidence-based health care decision-making requires judicious and critical comparison of all relevant competing interventions (Hoaglin et al., 2011; Jansen et al., 2011, 2014). Practice guidelines have also focused on the growing importance of prospective observational studies to inform health policy decisions (Berger et al., 2009, 2012, 2014). Additional good prac-tices on observational data may be found in AHRQ (2013), Alemayehu and Berger (2016), Berger and Doban (2016), and Vandenbroucke et al. (2007).

The ISPOR Good Research Practices Task Force published a guideline for conducting cost-effectiveness analysis (Ramsey et al., 2005, 2015). Topics include issues related to trial design, data element selection, database design and management, analysis, and reporting of results. Task force members rec-ommend that trials should be designed to evaluate effectiveness (rather than efficacy), should include clinical outcome measures, and should obtain health resource use information and health state utilities directly from study subjects.

Recently, the US Congress (2016) passed the landmark 21st Century Cures Act (H.R.34) as Public Law No. 114–255. Updating and expanding the so-called FDA Modernization Act (FDAMA) 114, Section 3027 of the 21st Cen-tury Cures Act provides guidance as to how a drug company may be able to communicate health care economic information with "a payer, formulary committee, or other similar entity with knowledge and expertise in the area of health care economic analysis, carrying out its responsibilities for the selection of drugs for coverage or reimbursement."

The FDA has issued a guidance for members of industry and the FDA administrative staff on the use of real-world evidence (RWE) to support regulatory decision-making for medical devices (FDA, 2017). In the Observational Medical Outcomes Partnership (OMOP, 2013), now Observational Health Data Sciences and Informatics (or OHDSI, 2017), it is underscored that the effective application of evidence-based medicine requires the close collaboration of all concerned stakeholders, including patients, health care providers, and payers.

In Europe, Makady et al. (2017) collected information on the RWE policies of six countries' health technology assessment agencies. These agencies included the following: (1) the Dental and Pharmaceutical Benefits Agency in Sweden, TLV; (2) the National Institute for Health and Care Excellence in the UK, NICE; (3) the Institute for Quality and Efficiency in Health Care in Germany, IQWiG; (4) the High Authority for Health in France, HAS; (5) the Italian Medicines Agency in Italy, AIFA; and (6) the National Healthcare Institute in the Netherlands, ZIN. However, how such evidence is used and leveraged is quite different, without uniformity across these various countries. Thus, across various countries, the use and acceptance of RWE for HTA purposes appears to be evolving.

8.4 Concluding Remarks

The health care industry can use HEOR data to promote public health through a better understanding of diseases and treatments. In addition, the industry can gain enhanced market access to bring affordable medicines to patients who need such treatments (AHRQ, 2013; Vandenbroucke et al., 2007). This brief chapter, which is intended to provide texture to other chapters in this book, highlights several aspects of best practices for use of HEOR data. In doing so, it may serve as a resource to leverage guidelines associated with the critical appraisal and conduct of HEOR studies. The available guidelines are important tools to help with the implementation of standard design and analytical approaches for HEOR.

References

Agency for Healthcare Research and Quality (AHRQ). 2013. *Developing a protocol for observational comparative effectiveness research: A user's guide.* http://www.effectivehealthcare.ahrq.gov/index.cfm/search-for-guides-reviews-and-reports/?pageaction=displayproduct&productid=1166&pcem=ra (accessed May 11, 2017).

Agency for Healthcare Research and Quality (AHRQ). 2017. *What is comparative effectiveness research?* http://www.effectivehealthcare.ahrq.gov/index.cfm/what-is-comparative-effectiveness-research1 (accessed May 11, 2017).

Alemayehu, D. and M. Berger. 2016. Big data: Transforming drug development and health policy decision making. *Health Serv Outcomes Res Methodol* 16:92–102.

Berger, M.L. and V. Doban. 2014. Big data, advanced analytics and the future of comparative effectiveness research. *J Comp Eff Res* 3:167–176.

Berger, M.L., Dreyer, N., Anderson, F. et al. 2012. Prospective observational studies to assess comparative effectiveness: The ISPOR good research practices task force report. *Value Health* 15:217–230.

Berger, M.L., Mamdani, M., Atkins, D. et al. 2009. Good research practices for comparative effectiveness research: Defining, reporting and interpreting non-randomized studies of treatment effects using secondary data sources: The ISPOR good research practices for retrospective database analysis task force report—Part I. *Value Health* 12:1044–1052.

Berger, M.L., Martin, B.C., Husereau, D. et al. 2014. A questionnaire to assess the relevance and credibility of observational studies to inform health care decision making: An ISPOR-AMCP-NPC good practice task force report. *Value Health* 17:143–156.

Calvert, M., Blazeby, J., Altman, D.G. et al. 2013. Reporting of patient-reported outcomes in randomized trials: The CONSORT PRO extension. *JAMA* 309:814–822.

Cappelleri, J.C., Deal, L.S., and C.D. Petrie. 2017. Editorial. Reflections on ISPOR's clinician-reported outcomes good measurement practice recommendations. *Value Health.* 20:15–17.

Cappelleri, J.C. and S.P. Spielberg. 2015. Advances in clinical outcome assessments. *Ther Innov Regul Sci* 49:780–782.

Caro, J.J., Eddy, D.M., Kan, H. et al. 2014. Questionnaire to assess relevance and credibility of modeling studies for informing health care decision making: An ISPOR-AMCP-NPC Good Practice Task Force report. *Value Health* 17:174–182.

Cohen, A., Goto, S., Schreiber, K. et al. 2015. Why do we need observational studies of everyday patients in the real-life setting? *Eur Heart J* 17:D2–D8.

Cox, E., Martin, B.C., Van Staa, T. et al. 2009. Good research practices for comparative effectiveness research: Approaches to mitigate bias and confounding in the design of nonrandomized studies of treatment effects using secondary data sources: The International Society for Pharmacoeconomics and Outcomes Research Good Research Practices for Retrospective Database Analysis Task Force Report—Part II. *Value Health* 12:1053–1061.

European Medicines Agency (EMA). 2005. Reflection paper on the regulatory guidance for us of health-related quality of life (HRQOL) measures in the evaluation of medicinal products. *Scientific Advice Working Party of CHMP.* http://www.ema.europa.eu/docs/en_GB/document_library/Regulatory_and_procedural_guideline/2009/10/WC500004201.pdf (accessed May 11, 2017).

European Medicines Agency (EMA), Scientific Advice Working Party of CHMP. 2014. *Qualification of novel methodologies for drug development: Guidance to applicants.* http://www.ema.europa.eu/ema/index.jsp?curl=pages/regulation/document_listing/document_listing_000319.jsp (accessed May 11, 2017).

Food and Drug Administration (FDA). 2009. Guidance for industry patient-reported outcome measures: Use in medical product development to support labeling claims. *U.S. Department of Health and Human Services.* http://www.fda.gov/downloads/Drugs/.../Guidances/UCM193282.pdf (accessed May 11, 2017).

Food and Drug Administration (FDA). 2014. Guidance for industry and FDA staff: Qualification process for drug development tools. *U.S. Department of Health and Human Services.* http://www.fda.gov/downloads/drugs/guidancecompliance-regulatoryinformation/guidances/ucm230597.pdf (accessed May 11, 2017).

Food and Drug Administration (FDA). 2016. *Clinical outcome assessment compendium.* http://www.fda.gov/Drugs/DevelopmentApprovalProcess/Developmen-tResources/ucm459231.htm (accessed May 15, 2017).

Food and Drug Administration (FDA). 2017. Use of real-world evidence to support regulatory decision-making for medical devices: Guidance for industry and food and drug administration staff. https://www.fda.gov/downloads/MedicalDevices/DeviceRegulationandGuidance/GuidanceDocuments/UCM513027.pdf. (accessed September 25, 2017).

Ford, I. and J. Norrie. 2016. Pragmatic trials. *N Engl J Med* 375:454–463.

Hoaglin, D.C., Hawkins, N., Jansen, J.P. et al. 2011. Conducting treatment-comparison and network-meta-analysis studies: Report of the ISPOR Task Force on Indirect Treatment Comparisons Good Research Practices—Part 2. *Value Health* 14:429–437

Holtorf, A.P., Brixner, D., Bellows, B. et al. 2012. Current and future use of HEOR data in healthcare decision-making in the United States and in emerging markets. *Am Health Drug Benefits* 5:428–438.

International Society for Pharmacoeconomics and Outcomes Research (ISPOR). 2016. Good practices for outcomes research reports. https://www.ispor.org/workpaper/practices_index.asp (accessed May 11, 2017).

Jansen, J.P., Fleurence, R., Devine, B. et al. 2011. Interpreting indirect treatment comparisons and network meta-analysis for health-care decision making: Report of the ISPOR Task Force on Indirect Treatment Comparisons Good Research Practices: Part 1. *Value Health* 14:417–428.

Jansen, J.P., Trikalinos, T., Cappelleri, J.C. et al. 2014. Indirect treatment comparison/network meta-analysis study questionnaire to assess relevance and credibility to inform health care decision making: An ISPOR-AMCP-NPC Good Practice Task Force Report. *Value Health* 17:157–173.

Johnson, M.L., Crown, W., Martin, B.C. et al. 2009. Good research practices for comparative effectiveness research: Analytic methods to improve causal inference from nonrandomized studies of treatment effects using secondary data sources: The ISPOR Good Research Practices for Retrospective Database Analysis Task Force Report--Part III. *Value Health* 12:1062–1073.

Makady, A., Ham, R.T., de Boer, A. et al. 2017. Policies for use of real-world data in health technology assessment (HTA): A comparative study of cix HTA agencies. *Value Health* 20:520–532.

Observational Health Data Sciences and Informatics (OHDSI). 2017. http://ohdsi.org (accessed May 11, 2017).

Observational Medical Outcomes Partnership (OMOP). 2013. http://omop.org (accessed May 11, 2017).

Patrick, D.L., Burke, L.B., Gwaltney, C.H. et al. 2011a. Content validity–Establishing and reporting the evidence in newly developed patient reported outcomes (PRO) instruments for medical product evaluation: ISPOR PRO good research practices task force report: Part 1–Eliciting concepts for a new PRO instrument. *Value Health* 14:967–977.

Patrick, D.L., Burke, L.B., Gwaltney, C.H. et al. 2011b. Content validity–Establishing and reporting the evidence in newly developed patient reported outcomes (PRO) instruments for medical product evaluation: ISPOR PRO good research practices task force report: Part 2–Assessing respondent understanding. *Value Health* 14:978–988.

Powers, J.H. III, Patrick, D.L., Walton, M.K. et al. 2017. Clinician-reported outcome (ClinRO) assessments of treatment benefit: Report of the ISPOR Clinical Outcome Assessment Emerging Good Practices Task Force. *Value Health* 20:2–14.

Ramsey, S., Willke, R., Briggs, A. et al. 2005. Good research practices for cost-effectiveness analysis alongside clinical trials: The ISPOR RCT-CEA Task Force Report. *Value Health* 8:521–533.

Ramsey, S.D, Willke, R.J., Glick H. et al. 2015. Cost-effectiveness analysis alongside clinical trials II—An ISPOR Good Research Practices Task Force Report. *Value Health* 18:161–172.

United States (U.S.) Congress. 2016. *H.R.34 - 21st Century Cures Act, 114th Congress (2015–2016).* Section 3037. https://www.congress.gov/114/bills/hr34/BILLS-114hr34enr.pdf (accessed May 11, 2017).

Vandenbroucke, J.P., von Elm, E., Altman, D.G. et. al. 2007. Strengthening the Reporting of Observational Studies in Epidemiology (STROBE): Explanation and elaboration. *Ann Inter Med* 147:W163–W194.

Willke, R.J. and C.D. Mullins. 2011. "Ten commandments" for conducting comparative effectiveness research using "real-world data." *J Manag Care Pharm* 17:S10–S15.

Zikopoulos, P.C., Eaton, C., deRoos, D. et al. 2012. Understanding Big Data: Analytics for Enterprise Class Hadoop and Streaming Data. New York, NY: McGraw Hill. https://www.ibm.com/developerworks/vn/library/contest/dw-freebooks/Tim_Hieu_Big_Data/Understanding_BigData.PDF (accessed May 11, 2017).

Index

A

ACER, *see* Average cost-effectiveness ratio
AMSTAR, *see* Assessment of Multiple Systematic Reviews
Assessment of Multiple Systematic Reviews (AMSTAR), 139
Average causal effect, 49
Average cost-effectiveness ratio (ACER), 87, 88, *see also* Incremental ACER

B

Bagging, 76, 77
Bayesian approach, 69, 100, 126–127
Biomarkers
    genomic, 152
    in personalized medicine, 155
Bivariate normal distribution, 96, 97
Body mass index, 59
Bootstrap aggregation, 76
Bucher method, 133

C

Case-control study, 50, 51
CAT, *see* Computer adaptive testing
Causal Effect, 49
Causal inference, confounding in, 48–50
CCT, *see* Classical test theory
CDM, *see* Common data model
CEA, *see* Cost-effectiveness analysis
CEAC, *see* Cost-effectiveness acceptability curve
CEAF, *see* Cost-effectiveness acceptability frontier
CEP, *see* Cost-effectiveness probability; Cost-effectiveness proportion
CER, *see* Comparative effectiveness research
CFA, *see* Confirmatory factor analysis
Chi-square test, 35, 132

Classical test theory (CCT), 19
Clinical Data Interchange Standards Consortium (CDISC)'s Healthcare Link project, 59
Clinical outcome assessments (COAs), 152, 178
    in personalized medicine, 152
    types, 157
Clinical trial phases, in drug development, 3–4
ClinicalTrials.gov, 3, 4
Cluster randomized trials (CRTs), 98
Cochrane Collaboration's risk of bias assessment, 137
Cochrane Consumer Network, 9
Cohort study, 2, 5, 50–51
Collaboratory Distributed Research Network (DRN), 7
Common data model (CDM), 59
Comparative effectiveness research (CER), 16, 179
Computer adaptive testing (CAT), 163
Conceptual framework, 18, 19
Concurrent validity, 24
Confirmatory factor analysis (CFA)
    assessing fit between model and data, 34–35
    exploratory factor analysis *vs.*, 31
    features, 35
    measurement model, 31–32
    parameters identification, 34
    real-life application, 35–36
    residual terms for endogenous variables, 33–34
    standard model for, 32–33
    standard *vs.* nonstandard model, 32
Confounding
    in causal inference, 48–50
    by indication, 48
Construct validity
    classical test theory and, 19
    definition of, 20

Content validity, 17–18
Convergent validity, 20
Cost-effectiveness acceptability curve
    (CEAC), 88–89
Cost-effectiveness acceptability frontier
    (CEAF), 89
Cost-effectiveness analysis (CEA),
    85–86, 98, 101
    decision analysis use in, 110
    example, 93–95
    probabilistic measures, 95–99
    statistical inference for, 89–90
    measures, 87–88
    using Markov model, 111–115
Cost-effectiveness probability (CEP),
    95–99
Cost-effectiveness proportion (CEP),
    95, 96
Cost-effectiveness ratio, 87, 88, *see also*
    Average cost-effectiveness
    ratio; Incremental cost-
    effectiveness ratio
Counterfactual causality, 49
Criterion validity, 24, 42
Cronbach's alpha, 41–42
Cross-sectional studies, 52
Cross-validation (CV), 71
CRTs, *see* Cluster randomized trials
CTT, 19, 36
Cumulative meta-analysis, 127–128, 144
CV, *see* Cross-validation
CVMSE, 71
CVMSR, 71

D

Data incorporation, 98–99
    effectiveness data from observational
        studies, 101
    indirect comparisons, 99–101
    issues with cost data, 102–104
Data management tools, 58
Data sharing, 6, 7
Data sources, and evidence hierarchy, 2–3
Data warehousing, 58–59
Decision analysis, 104–105
    CEA example using Markov model,
        111–115

in cost-effectiveness analysis, 110
    decision tree, 107
    Markov models, 108–110
    outcome measures, 105–106
    sensitivity analysis, 110–111
    steps in, 105
Decision-making, 87, 110, 154
Design options, 50–52
Diagnostic biomarkers, 155
Digital revolution, 1
Discriminant validity, 20
Divergent validity, 20
Drug development, clinical trial phases
    in, 3–4
Duan's approach, 103

E

Economic evaluation, in precision
    medicine, 153–154
ED, *see* Erectile dysfunction
EDWs, *see* Enterprise data warehouses
EFA, *see* Exploratory factor analysis
EHS, *see* Erection Hardness Score
Electronic health records (EHRs), 5, 6
EMAX model, 74
Enhancing the quality and transparency
    of health research (EQUATOR)
    Network, 3
Enterprise data warehouses (EDWs),
    58
EQ-5D, 106
Erectile dysfunction (ED)
    clinical diagnosis of, 40–42
    levels of, 21–22
Erection Hardness Score (EHS), 40
Evidence hierarchy, data sources and, 2
Evidence pyramid, 2
Exchangeability, 49, 133
Exploratory factor analysis (EFA),
    24–25
    assumptions, 29–30
    factor rotation, 28
    model, 25–27
    number of factors, 27–28
    real-life application, 30
    role of, 25

sample size, 28–29
vs. confirmatory factor analysis, 31

F

Factor-analytic model, 29
False discovery rate (FDR), 156, *see also*
    Positive false discovery rate
FDA, *see* Food and Drug
    Administration
FDA Modernization Act (FDAMA), 179
FDR, *see* False discovery rate
Fieller's method, 89
Fixed effects model, 10, 130–131
Food and Drug Administration (FDA),
    4, 180

G

Generalized linear mixed effects, 162
Generalized linear model (GLM), 103
Generalized pivotal quantity (GPQ),
    91–93
Genome-wide association study, 155
Genomic analysis, statistical issues in,
    155–156
Genomic biomarkers, 152, 153
GLMNET, 73
Grades of Recommendation,
    Assessment, Development
    and Evaluation (GRADE), 139,
    140
Greedy matching, 53

H

hd-PS, *see* High-dimensional propensity
    score
Health care industry, 1
Health economics and outcomes
    research (HEOR), 177
    applications of predictive models in,
        79–80
    exploratory analysis and
        premodeling strategies, 70–72
    high-dimensional data analyzing,
        77–78

linear predictive models, 72–74
nonlinear predictive models, 74–76
practices for, 179–180
software, 78
tree-based methods, 76–77
Health Information Technology for
    Economic and Clinical Health
    (HITECH) Act, 61
Health states, 108
Health technology assessment (HTA),
    154, 166–167
HEOR, *see* Health economics and
    outcomes research
Heterogeneity, 124, 131, 132, 136
Hierarchical model, 156
High-dimensional propensity score
    (hd-PS), 55
Himmelfarb Health Sciences Library, 2
HITECH Act, *see* Health Information
    Technology for Economic and
    Clinical Health Act
Homogeneity, evaluation of, 132
HTA, see Health technology
    assessment

I

ICC, *see* Intraclass correlation
    coefficient
ICER, *see* Incremental cost-effectiveness
    ratio
IIEF, *see* International Index of Erectile
    Function
INB, *see* Incremental net benefit
Incremental ACER (ΔACER), 88, 90, 93
Incremental cost-effectiveness ratio
    (ICER), 87, 88
    generalized pivotal quantity for, 92
    parameter, 91
Incremental net benefit (INB), 87
Individual randomized trials, 137
Instrumental variables (IVs), 55–56
Internal reliability, 39
    consistency, 41–42
International Classification of Diseases,
    Ninth Revision, Clinical
    Modification (ICD-9-CM), 60

International Classification of Diseases, Tenth Revision, Clinical Modification (ICD-10-CM), 60
International Index of Erectile Function (IIEF), 24, 40
International Society for Pharmacoeconomics and Outcomes Research (ISPOR), 98, 140, 178
   good research practices task force, 179
Intraclass correlation coefficient (ICC), 40
Inverse probability of treatment weighting (IPTW) method, 54
IRT, *see* Item response theory
ISPOR, *see* International Society for Pharmacoeconomics and Outcomes Research
Item-level discriminant validity, 20
Item response theory (IRT), 36, 162–163
IVs, *see* Instrumental variables

K

Kappa coefficient, 40
Kappa statistic, 40
k-nearest neighbors (*k*-NN), 75, 79
Known-groups validity, 21–23

L

LASSO, *see* Least absolute shrinkage and selection operator
LDA, *see* Linear discriminant analysis
Least absolute shrinkage and selection operator (LASSO), 52, 73, 77, 79, 81
Least squares regression, 72
Linear discriminant analysis (LDA), 74
Linear mixed-effect models, 161
Linear predictive models, 72–74
Linear regression model, 72, 154

M

Machine learning literature, developments in, 57
MAER-Net, *see* Meta-Analysis of Economics Research-Network

MAR, *see* Missing at random
Markov cohort simulation, 109
Markov models, decision analysis, 108–110
MARS, *see* Multivariate adaptive regression splines
Matching mechanism, 51
MAUS, *see* Multi-attribute utility scale
Maximum likelihood principle, 69
Mean squared error (MSE), 71–72
Medical evidence generation, 1
Medical practice, interruptions in, 55
Meta-analysis, *see also* Network meta-analysis
   Bayesian framework, 126
   conduct and reporting of, 136–140
   cumulative, 127
   defined, 143
   frequentist framework, 125
   random effects, 140
   model validation, 131
   systematic reviews and, 9–10
Meta-Analysis of Economics Research-Network (MAER-Net), 138
Meta-Analysis of Observational Studies in Epidemiology (MOOSE), 138
Meta-regression, 135–136
Minnesota Nicotine Withdrawal Scale (MNWS), 30, 35
Misinformation chaos, 156
Missing at random (MAR), 165
Missing not at random (MNAR), 165
MNWS, *see* Minnesota Nicotine Withdrawal Scale
MOOSE, *see* Meta-Analysis of Observational Studies in Epidemiology
MSE, *see* Mean squared error
Multi-attribute utility scale (MAUS), 106
Multivariate adaptive regression splines (MARS), 78

N

National Drug Code (NDC) scheme, 60
National Institute of Health (NIH), 4

Collaboratory Distributed Research Network, 7
Patient-Reported Outcomes Measurement Information System, 162
NDC scheme, *see* National Drug Code scheme
Nearest-neighbor matching, 53
Negation, 132
Nested case-control study, 51
Network meta-analysis (NMA), 128–129, 139
  consistency and transitivity, 133
  fixed effects model, 130–131
  homogeneity, 132
  random effects model, 131
Neural networks, 75
NIH, *see* National Institute of Health
NMA, *see* Network meta-analysis
Nonlinear predictive models, 74–76
Nonparametric regression models, 78

O

Observational data, 1
Observational Medical Outcomes Partnership (OMOP), 59
Observational studies, 5–7
  analysis and reporting of, 61
  methodological challenges, 7
OLS, *see* Ordinary least squares
OMOP, *see* Observational Medical Outcomes Partnership
Operational considerations, 57–62
  computing and data visualization, 60
  data standards, 59–60
  data warehousing and processing, 58–59
  security and privacy, 60–62
Optimal decision-making, 154
Optimal matching, 53
Optimal treatment strategy, 154
Ordinary least squares (OLS), 102

P

Partial least squares (PLS), 72–74
Patient-Centered Outcomes Research Institute (PCORI), 16

Patient-reported outcome (PRO), 8–9, 152
  guidelines on, 178
  measurement, 16–20
  missing data with, 163–165
  in precision medicine, 157–159
    instrument development and validation, 159–160
    interpretation, 165–166
    item response theory, 162–163
    longitudinal models, 161–162
Patient-Reported Outcomes Measurement Information System (PROMIS), 162
PCA, *see* Principal component analysis
PCORI, *see* Patient-Centered Outcomes Research Institute
Personalized medicine, 151
  biomarkers in, 155
  component of, 154
  recent advances in, 152
Person-item maps, 36–39, 38
Placebo group, 3
Positive false discovery rate (pFDR), 156
PPR, *see* Projection pursuit regression
Pragmatic trials, 7–8
Precision medicine, 151
  clinical practice, 168–169
  economic evaluation in, 153–154
  ethical issues, 168
  health technology assessment, 166–167
  patient-reported outcomes in, 157–159
    instrument development and validation, 159–160
    interpretation, 165–166
    item response theory, 162–163
    longitudinal models, 161–162
    missing data with, 163–165
  regulatory issues, 167–168
Predictive biomarkers, 155
Predictive model, 69
  applications, 79–80
  linear, 72–74
  nonlinear, 74–76
Predictive validity, 24
Predisposition biomarkers, 155

Preferred Reporting Items of
Systematic reviews and Meta-
Analyses (PRISMA) statement,
137–138
Principal component analysis (PCA),
25, 72, 79
Prognostic biomarkers, 155
Projection pursuit regression (PPR), 78
PROMIS, *see* Patient-Reported Outcomes
Measurement Information
System
Propensity score (PS), 49, 52–55
Publication bias, model validation,
134–135

Q

quality-adjusted life-years (QALYs), 106,
107, 112, 115
Quality of life (QOL), 158, 159
Quality of reporting of meta-analyses
(QUOROM), 137

R

Random effects meta-analysis, 140
Random effects model, 131
Randomized controlled trials (RCTs),
7, 8, 47–48
individual, 2
and observational data, 6
patient-reported outcomes in, 178
Rasch model, 36, 37
Real world data (RWD), 5–7
in application, 62–63
in patient populations, 8
Real-world evidence (RWE), 5
Regression adjustment analysis
methods, 52
Regulatory agencies, 1
Reliability
internal, *see* Internal reliability
repeatability, 39–41, 160
test-retest, 39
Repeatability reliability, 39–41, 160
Responsiveness, 23
Retrospective cohort databases, 5
Ridge regression, 72

Root-Mean-Square Error of
Approximation (RMSEA), 34
RWD, *see* Real world data
RWE, *see* Real-world evidence

S

SAS®, 140–142
Saturation, 18
Scree test, 27
Self-controlled cohort design, 51
Self-controlled study, 51
Self-Esteem And Relationship (SEAR)
questionnaire, 21, 22
SEM, *see* Structural equation model
Semi-Markov microsimulation models,
108
Sensitivity, 22, 23
analysis, 56, 110–111
Sentinel CDM (SCDM), 59
Software, 78
Stable-unit-treatment assumption, 49
Standard gamble approach, 106
Statistical methods, 69
Strengthening the Reporting of
Observational studies in
Epidemiology (STROBE), 62
Structural equation model (SEM),
31, 56–57
Subgroup analysis, 156–157
Sufficient-component-cause model, 50
Support vector machines (SVMs),
75, 79, 81
Systematic reviews, 9–10

T

Technical support documents (TSD), 139
Test-retest reliability, 39
Thrombolytic drug, 128
Time-dependent covariates, 57
Transitivity, 133
Treatment-by-study interaction, 132
Tree-based methods, 76–77
TSD, *see* Technical support documents
Tunnel states, 109
21st Century Cures Act, 179
Two-stage least-squares method, 56

U

Univariate techniques, 154
Unrelated means effects (UME), 133

V

Validity
    concurrent, 24
    construct, *see* Construct validity
    content, 17–18
    convergent, 20
    criterion, 24
    known-groups, 21–23
    predictive, 24
    types, 17

W

Wavelet regression, 78
Weighted kappa, 40

Printed in the United States
by Baker & Taylor Publisher Services